A GUIDE TO METHODOLOGY IN ERGONOMICS

A GUIDE TO METHODOLOGY IN ERGONOMICS

Designing for human use

Neville A. Stanton and Mark S. Young

London and New York

First published 1999 by Taylor & Francis
11 New Fetter Lane, London EC4P 4EE

Simultaneously published in the USA and Canada
by Routledge
29 West 35th Street, New York, NY 10001

Taylor & Francis is an imprint of the Taylor & Francis Group

Typeset in Times by J&L Composition Ltd, Filey, North Yorkshire
Printed and bound in Great Britain by T.J. International Ltd, Padstow, UK

British Library Cataloguing in Publication Data
A catalogue record for this book is available from the British Library

Library of Congress Cataloging in Publication Data
Stanton, Neville
A guide to methodology in ergonomics: designing for human use/
Neville A. Stanton & Mark S. Young
p. cm
Includes bibliographical references and index.
1. Human engineering. I. Young, Mark S. II. Title.
TA166.S73 1999
620.8′2—dc21   99–25464
CIP
ISBN 0–7484–0703–0

For Maggie, Joshua and Jemima
Neville

For Mum, Dad and Bruv
Mark

CONTENTS

FOREWORD

The availability of reliable and valid ergonomics methodologies enables the effective analysis and design of tasks to be performed by humans. By using these, the system's operational effectiveness should improve through reduced error rates, faster learning and performance times, and higher satisfaction in task performance.

Ergonomics methodologies encompasses physical, psychophysical, social and psychological methods. They also include specific methodologies relating to ergonomics audits and the use of computer software for ergonomics modelling. Physical methodologies deal with such topics as anthropometric measurements, physiological expenditure of energy and muscular exertions. Psychophysical methodologies deal with such topics as physiological measures of psychological activities (e.g. EEG, EOG, sinus arrhythmia, and breathing rates). Social ergonomics methodologies deal with the analysis, composition and design of group work and the whole field of macroergonomics.

The purpose of this book is to concentrate on a select but widely utilized group of psychological ergonomics methods, hence its publication is very timely. It provides practitioners with valuable insight into how, where and why each of the methodologies cited in the book has its unique utility and advantage. The step-by-step process provided with each methodology and the abridged nature of the book will make this a requirement for every ergonomics practitioner's bookshelf.

Gavriel Salvendy
NEC Professor of Industrial Engineering
Purdue University
March 1999

PREFACE

This book began as a two-year research project into ergonomic methods funded by the Engineering and Physical Sciences Research Council under the LINK Transport Infrastructure and Operations Programme. It has five objectives:

- To review product usability methods.
- To determine which methods are most appropriate.
- To provide a methodology for selecting methods.
- To illustrate how the methods work.
- To validate the methods.

Although the methods were selected for their potential applicability to in-car devices, we believe the approach we have developed can be applied more widely. The final output of the project was a manual of ergonomic methods, on which this book is based. We also developed some software to accompany the manual. Together these tools can help the designer decide which ergonomic methods are most appropriate at any point in product design. The emphasis has been upon applying the methods early in the design process and quantifying the benefit to the users of the methods in an explicit manner. We feel this book could serve as a basis for training people in ergonomic methods. Initial reactions from designers have been very positive. We have started to publish the results from the research project; details may be found in Stanton and Young (1997a, 1997b, 1997c, 1998a, 1998b, 1999a, 1999b).

An accompanying software application helps you to decide online which methods to use. This is avaliable from the authors for a modest sum to cover administration, postage and packing, and the price of a disk (please specify Mac or PC).

Neville Stanton
Professor of Design
Department of Design
Brunel University
Runnymede Campus
Coopers Hill Lane
Egham, Surrey TW20 0JZ
United Kingdom

Mark Young
Engineering Psychology Research Group
Department of Psychology
University of Southampton
Southampton SO17 1BJ
United Kingdom

ACKNOWLEDGEMENTS

We would like to thank all those that have been involved in the project. Howard Wyborn (our programme manager) for staying the course, Mike Bradley and Bharat Patel (our industrial contacts) for having faith, all the participants, Dr Tony Roberts for helping us to unravel the statistical knots, Luke Hacker (Taylor & Francis) for encouraging us to publish the research in this form, the anonymous project reviewers who looked favourably upon this project, Professor Gavriel Salvendy for writing the foreword, and last but not least, the EPSRC and LINK for providing the research funds that made the project possible.

HOW TO USE THIS BOOK

This book is designed for expert ergonomists (i.e. persons with postgraduate training in ergonomics) to help them conduct an ergonomics analysis of a product in development. If it is used by someone less qualified, then it needs to be in conjunction with training in ergonomics. This training should be provided by someone with both experience and expertise in ergonomics methods as well as suitable knowledge and experience of training. We would recommend contacting the Ergonomics and/or Human Factors Society in your country. A list of recent contacts was published by Stanton (1998). The book details the execution of 12 ergonomic methods. You may read the book in different ways – we hope to have accommodated for the complete novice (with the appropriate training) as well as the expert ergonomist.

The book is the result of two years' research into the development of a methodology for the safer operation of in-car devices funded by the EPSRC under the LINK Transport Infrastructure and Operations Programme, hence the preoccupation with the analysis of radio-cassettes in the examples provided. We believe that the methods presented in the book can be applied to the design of all manner of devices and interfaces. It is argued that ergonomics has a practical role in this endeavour, on the basis that the methods illustrated here should help in several ways to improve design performance:

- They reduce device interaction time.
- They reduce user errors.
- They improve user satisfaction.
- They improve device usability.

There are four major sections to this book. Section 1 is the introduction. This contains an overview, providing some background on what ergonomic methods do, where they fit into the design process and how to select a method appropriate for your purpose.

Section 2 describes the methods themselves. For each method there is an overview, instructions on how to carry out an analysis, a mini bibliography, pros and cons, examples or an example, and a flowchart. The pros and cons are presented under four headings:

- *Reliability/validity*: based on empirical data, reliability and validity measure how well the method predicts what it is supposed to, and how stable it is between users and over time.

Ergonomics texts: an overview

Editors or authors	Title	Edited or authored	Date (edition)	Pages	Number of methods	Domain	Application of methods	Number of validation studies	Number of training studies
Diaper	Task analysis in HCI	edited	1989 (1st)	258	6	specific	specific	0	0
Kirwan and Ainsworth	A guide to task analysis	edited	1992 (1st)	417	23	generic	generic	0	0
Kirwan	A guide to practical HRA	authored	1994	592	28	generic	specific	2	0
Corlett and Clarke	Ergonomics of workspace and machines	edited	1995 (2nd)	128	6	generic	generic	0	0
Wilson and Corlett	Evaluation of human work	edited	1995 (2nd)	1134	48	generic	generic	1	0
Jordan et al.	Usability evaluation in industry	edited	1996 (1st)	252	20	generic	generic	2	0
Salvendy	Handbook of human factors and ergonomics	edited	1997 (2nd)	2137	over 100	generic	generic	1	0
Stanton	Human factors in consumer products	edited	1998 (1st)	287	27	specific	specific	0	0

- Resources: again using data derived by experiment, resources are mainly concerned with how long the method takes to train, practice and apply; how many people are required to carry it out; and if there are any prerequisites for its use.
- *Usability*: trained analysts used each of the methods and rated them on seven scales of usability. These data are used to assess the method on perceived ease of use.
- *Efficacy*: this is a more qualitative analysis; it specifies the stages during design when the method can be used, the form of the product required for a meaningful analysis, the type of output the method provides, whether access to end-users should be appropriate, etc.

The greater the number of stars (*) assigned to any criterion, the better its performance. The examples are based on two simple car radio designs, illustrated on pages xviii and xix. It might be helpful to photocopy these figures so they can be laid alongside each method review.

Section 3 looks at the data from studies on training, reliability and validity. From our research, we have found almost nothing in the open literature on these topics (Stanton and Young, 1998a). We see one of the aims of this book as to stimulate more research effort in this area. Although our data cannot be thought of as definitive, they do offer some basis for comparing methods. We would encourage more training and cross-validation studies.

Utility analysis is the subject of Section 4. It presents an equation which enables the analyst to calculate approximate financial benefits of using each method. This is to assist in the selection of methods. Although we would accept that the reason for choosing a method should not be purely financial (perhaps the method may not look particularly cost-effective, but it could still enhance the usability of the product and therefore the reputation of the manufacturer), it does help by increasing the objectivity of selecting methods. There is a full bibliography at the end of the book.

We hope the book will encourage designers to be more adventurous when choosing methods. Like others who have undertaken similar surveys, it is our experience that people tend to stick with a few of their favourite approaches no matter what task they are undertaking. We propose that people should choose methods fit for the task, and this requires designers to be careful in specifying their requirements at the outset.

If you are a novice, read all of the introduction to familiarise yourself with the task at hand before embarking on the analysis. The utility analysis is optional, depending on your requirements. Once you have selected a method or methods (you may want to use more than one if you have multiple requirements) then turn to Section 2 to learn more about it. If you are still unsure after reading this, we recommend using the key references for your selected method and seeking training from qualified personnel.

At the other end of the scale, the expert ergonomist probably only need use this book as a reference source. Even the most ardent practitioner may have trouble remembering how to use every method, so Section 2 will undoubtedly be a useful memory aid. You will probably find the utility analysis equation of interest, as well as our methodology for selecting an appropriate method.

A book can only cover so much, and as with all ergonomics methods, there is an element of craft skill that can only be acquired through practice. We strongly recommend the novice to begin with a relatively small, self-contained project and build up to larger, more complex projects. Always refer to an expert while learning how to apply the methods.

Here are some sources of information on ergonomics and ergonomic methods that we recommend:

- Corlett, E. N. and Clarke, T. S. (1995) *The Ergonomics of Workspaces and Machines*, 2nd edn, London: Taylor & Francis.
  Contains information on physical ergonomics and guidelines for design as well as information on six methods in some 128 pages. Methods include layout analysis, checklists, hierarchical task analysis, observation, control display design and anthropometrics. A4 layout in a practical format.

- Diaper, D. (1989) *Task Analysis in Human Computer Interaction*, Chichester: Ellis Horwood.
  Contains detailed accounts of six task analysis methods, specific to HCI, but could be more widely applied. Methods include hierarchical task analysis, task-action grammar, task analysis for knowledge descriptions, task knowledge structures and task object modelling. Examples throughout the 258 pages.

- Jordan, P. W., Thomas, B., Weerdmeester, B. A. and McClelland, I. L. (1996) *Usability Evaluation in Industry*, London: Taylor & Francis.
  Contains brief examples of some 20 methods in 252 pages. Methods include observation, participation, focus groups, usability testing, questionnaires, interviews, think-aloud protocols, hierarchical task analysis, task analysis for error identification, field studies, repertory grids and checklists. Benefits from the practical advice of practising ergonomists.

- Kirwan, B. (1994) *A Guide to Practical Human Reliability Assessment*, London: Taylor & Francis.
  Contains details of 28 human reliability methods applied to safety-critical industries, although some of these methods could be applied to product design. Methods include absolute probability judgement, confusion matrix, generic error modelling, human error assessment and reduction technique, influence modelling and assessment system, potential human error causes analysis, systematic human error reduction and prediction approach and success likelihood index method; 592 pages.

- Kirwan, B. and Ainsworth, L. (1992) *A Guide to Task Analysis*, London: Taylor & Francis.
  Contains 25 methods in 417 pages. Methods include activity sampling, barrier analysis, networking techniques, critical incident technique, checklists, failure mode and effects analysis, hazard and operability study, hierarchical task analysis, link analysis, observation, interviews, questionnaires and verbal protocols. A wide range of methods applied to large-scale system design (e.g. power stations) with detailed examples in the appendix.

- Norman, D. A. (1988) *The Psychology of Everyday Things*, New York: Basic Books. A good, non-technical, introduction to ergonomics and its application to the design of consumer products. Many examples of good and poor design with the underlying psychological theory of why they work or not. Fun to read!

- Salvendy, G. (1997) *Handbook of Human Factors and Ergonomics*, New York: Wiley. A truly mammoth textbook containing information on well over 100 methods in 2137 pages. This is probably the most comprehensive ergonomics text which covers just about every conceivable topic. In general, the details on each of the methods are brief, but pointers are given for further information.

- Stanton, N. A. (1998) *Human Factors in Consumer Products*, London: Taylor & Francis.
  Contains a range of products and techniques brought to bear on their design, including checklists, hierarchical task analysis, observation, interviews, error prediction, questionnaires, guidelines, focus groups, simulations and user trials. Plenty of examples are given throughout the book; 27 methods in 287 pages.

- Wilson, J. R. and Corlett, E. N. (1995) *Evaluation of Human Work*, 2nd edn, London: Taylor & Francis.
  Contains 48 methods in 1134 pages, under the generic headings of direct and indirect observation, modelling, task analysis, simulation, prototyping, subjective assessment, human reliability analysis, accident reporting, workload assessment and guidelines. It provides a comprehensive and authoritative overview of ergonomics.

We have analysed the methods reported in the eight texts cited above in order to determine how many methods are cited, and the emphasis is given to generality of the domain of application, as well as looking for evidence to help us answer the four questions we raise regarding ergonomic methods. This analysis, albeit subjective (see table p. xiii), shows that all but one of these texts are multi-authored and all but one were produced in this decade, although three are second editions. The number of methods contained within the texts ranges from 6 to over 100. Most of the texts are general in nature. Four of the texts contain validation studies, but these are sparse and only apply to a few of the methods mentioned. Finally, none of the texts contain any description of studies that relate to the acquisition of the method or, apart from Kirwan (1994), the relative merits of one method over another. This book aims to redress this position by providing data on training people to use the methods as well as data on their reliability and validity.

We recommend consulting journals such as *Ergonomics, Human Factors, Applied Ergonomics, Behaviour and Information Technology, Ergonomics in Design, Interacting with Computers, International Journal of Cognitive Ergonomics, Interational Journal of Human–Computer Interaction* and *Safety Science*. Conferences are also a useful way of keeping up to date with developments, such as the Ergonomics Society Annual Conference in the United Kingdom and the Human Factors and Ergonomics Society Annual Conference in the United States.

FORD 7000 RDS-EON

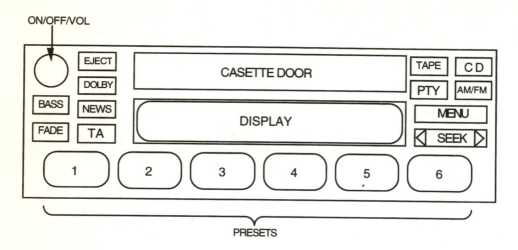

This is a schematic diagram of the Ford radio-cassette referred to in the case studies. It is an advanced device; among its functions are RDS, automatic music search and programme type tuning. It is about twice the height of a standard radio, and apart from the on/off/volume control, all the controls are buttons. Some of the controls may need further elaboration.

- ON/OFF/VOL: combined push-button/knob control, radio is on when in the 'out' position. Turn clockwise to increase volume.
- BASS/TREB: moded function. To adjust bass, push once, then use volume knob. To adjust treble, push twice, then use volume knob.
- FADE/BAL: moded function. To adjust fade, push once, then use volume knob. To adjust balance, push twice, then use volume knob.
- NEWS and TA: RDS functions. Push once to activate the News and Traffic Announcement interrupt functions respectively.
- TAPE: switches into Tape mode from Radio or CD mode. Serves as auto-reverse button when in Tape mode.
- PTY: Programme Type tuning. Push once, then use volume knob to select programme type. Use seek buttons to find programme type.
- CD: switches into CD mode from Tape or Radio mode.
- AM/FM: switches into Radio mode from Tape or CD mode. Serves as waveband selector when in Radio mode.
- MENU: many of the advanced functions are in the Menu. Push to scroll through the functions, then typically use the Seek buttons to adjust each function.
- SEEK: multi-functional. Use to tune radio, fast-forward/rewind cassette, and adjust many of the Menu functions.

SHARP RG-F832E

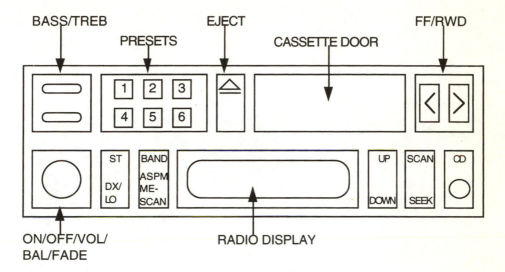

BASS/TREB   EJECT   FF/RWD

PRESETS   CASSETTE DOOR

ON/OFF/VOL/   RADIO DISPLAY
BAL/FADE

This is a schematic diagram of the Sharp radio-cassette referred to in the examples. It is a rather standard radio and has no RDS facilities. Some of the controls may need further elaboration:

- ON/OFF: this is a knob-twist control; turn clockwise for on, then further to increase volume. Push and turn to adjust fade; a collar adjusts balance.
- ST: push the top of this button to toggle stereo/mono radio reception.
- DX/LO: toggles local or distance reception when scanning for radio stations.
- BAND: switches between wavebands.
- ASPM ME-SCAN: scans the preset stations.
- UP/DOWN: manual radio tuning.
- SCAN: scans the current waveband for radio signals; continues until interrupted by user.
- SEEK: looks for next radio signal on current waveband and locks onto it.
- CD: CD/auxiliary input socket.
- BASS/TREB: sliding controls for bass and treble.

Section 1

INTRODUCTION

This book is designed to assist a novice user of ergonomic methods in learning and then carrying out their own analysis on a product. It also supports the expert as an aide-memoire. It is probably best used as a reference source or supplement to a training course in ergonomic methods.

What are ergonomic methods?

Ergonomic methods are designed to improve product design by understanding or predicting human interaction with those devices. Different methods tap different aspects of this interaction. An account of the interaction, based upon Norman's (1988) seven-stage model of human action, is illustrated in Figure 1.1. We accept this is a grossly simplified account of human activity (it does not describe emotional, personality, gender, cultural and social factors), but it does help to explain the different aspects of human performance addressed by the methods under examination in this book.

As Figure 1.1 shows, purposeful human action begins with the need to satisfy a particular goal, such as to follow England's progress in the test match (i.e. listen to a cricket match). This in turn leads to the formulation of intentions, such as to turn the car radio on (which will draw upon long-term memory). The intention is translated into a sequence of actions, which are finally executed. Action execution will comprise taking a hand off the steering wheel and moving it towards the radio, homing in on the power button, pressing the power button, homing in on the volume knob, turning the volume knob to the desired level. All of these actions have associated feedback loops, to let our cricket fan know whether they were successful or not. This requires the individual to perceive the state of the world (e.g. to receive information from the world in the form of tactile, auditory and visual input) and to interpret that perception: Have changes in the world occurred, e.g. is the radio on? Finally the evaluation of the interpretations is related to the goal: Am I listening to the cricket? Again this draws on long-term memory. Some of this activity will occur in conscious working memory (distinct from the longer-term store), whereas other activity may be largely automated; it occurs without conscious attention or with very little conscious attention. All manner of things could go wrong: the radio could be tuned to another radio station, or it could be faulty; or the car could be in a tunnel and therefore not picking up any radio stations. This would lead to changes in the intentions, and the activity cycle would begin again.

Each of the methods considered in this book focuses on different aspects of human performance. The methods may be broadly classified as quantitative or qualitative approaches. All of the methods make predictions about the user, the device, or the user and the device. The quantitative methods predict speed of performance (e.g. keystroke level model KLM), errors (e.g. predictive human error analysis (PHEA) and task analysis for error identification (TAFEI)) and speed and errors (e.g. observations). The qualitative methods predict user satisfaction (e.g. questionnaires and repertory grids), device optimisation (e.g. checklists, link analysis and layout analysis) or user and device interaction (e.g. heuristics, hierarchical task analysis (HTA) and interviews). Figure 1.2 indicates the extent to which the methods address the user–device activity cycle.

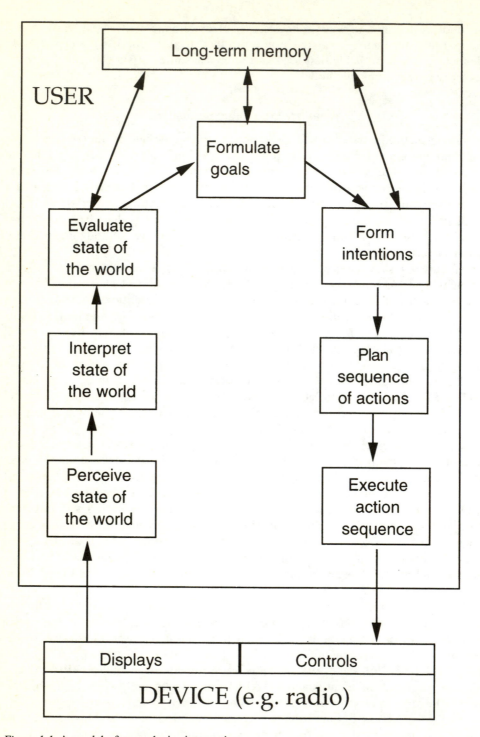

Figure 1.1 A model of user–device interaction.

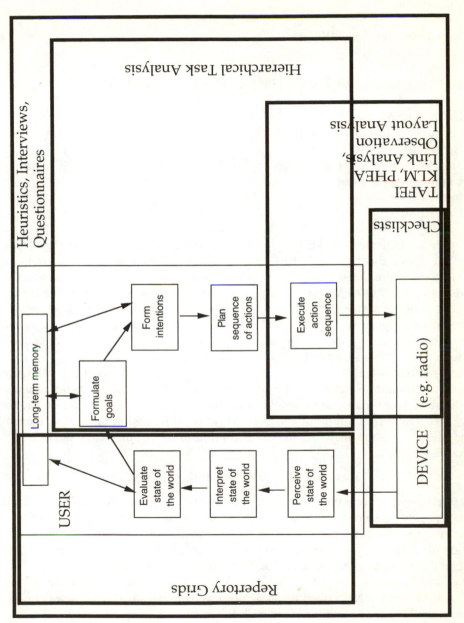

Figure 1.2 How ergonomic methods address the user–device activity cycle.

Twelve methods are described, starting with the most effective and working through to the least effective:

- Keystroke level model (KLM)
- Link analysis
- Checklists
- Predictive human error analysis (PHEA), also called SHERPA
- Observation
- Questionnaires
- Hierarchical task analysis (HTA)
- Repertory grids
- Task analysis for error identification (TAFEI)
- Layout analysis
- Interviews
- Heuristics

As Figure 1.2 indicates, checklists are predominantly concerned with the characteristics of the device. By contrast, TAFEI, PHEA, KLM, link analysis, layout analysis and observation capture the interaction between the execution of actions and the resultant device changes. Hierarchical task analysis represents the goals and intentions, plans and actions of the device user, whereas repertory grids represent the user's perceptions of the device in terms of psychological constructs. Finally, heuristics, interviews and questionnaires attempt to capture the whole essence of user–device interaction, although not all of this may be explicitly represented.

These methods were selected from a pool of 30 and have been subject to a validation study to determine how effective they are when used to assess a car radio. Some of them require more than one person in order to conduct the analysis, and some are better executed using a sample of end-users. However, it is possible to conduct them all in the laboratory using analysts with representations only, not the actual devices.

How do I choose which method to use?

Some methods are only appropriate to use at certain points in the design cycle, some take longer than others, some depend on two people being available, and all provide different forms of output. So at the end of the day, your choice is virtually made for you!

The following tables summarise the criteria for using all the methods, and working through them should help you to decide. A software version is available.

The first question to ask yourself is: At which stage of the design cycle are you? As ergonomists, we have generalised the design process into five major categories at which different methods may be applied (Figure 1.3):

- *Concept* maps onto the prestrategic intent stage; depending upon the product, this stage may be some years before the product launch.
- *Design* seems to relate best to the months until program approval.

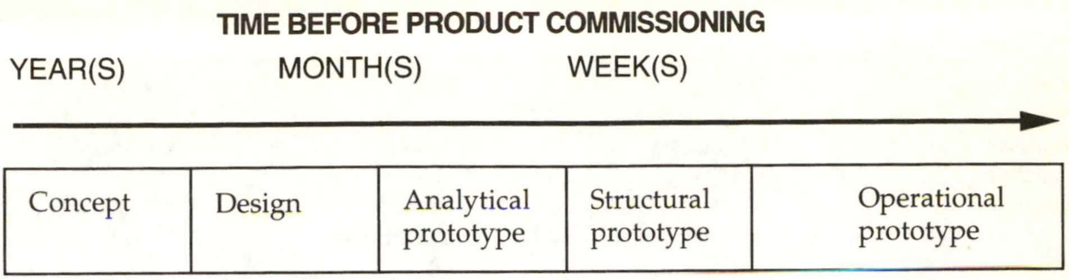

TIME BEFORE PRODUCT COMMISSIONING

YEAR(S)   MONTH(S)   WEEK(S)

Concept	Design	Analytical prototype	Structural prototype	Operational prototype

Figure 1.3 Principal stages of design.

- *Prototype* can be seen as the stage up to product readiness and analytical sign-off; this is normally less than a year from intended product launch (we have seperated this into analytical and structural phases to correspond with computer aided design (CAD) and hard-built prototypes).
- *Operational* may be related to any time from launch readiness to a few months before actual commissioning.

We believe that the methods may have the greatest impact at the prototyping stage, particularly considering one of the key design stages – analytic prototyping. Although in the past, it may have been costly to alter design at structural prototyping, and perhaps even impossible, computer-aided design has made the retooling much simpler. And it may even allow alternative designs to be compared at this stage. These ideas have yet to be proven in practice; however, given the nature of most ergonomic methods, it would seem most sensible to apply them at the analytic prototyping stage.

We have identified three main forms of analytical prototyping for human interfaces: functional analysis (the range of functions the device supports), scenario analysis (performing particular sequences of activities) and structural analysis (non-destructive testing from a user-centred perspective). We have classified the methods in this manual into each of these types. We hope this helps to highlight the different contributions each of the methods makes to analytical prototyping.

Functional analysis	*Scenario analysis*	*Structural analysis*
Interviews	Link analysis	KLM
Questionnaires	Layout analysis	PHEA
Checklists	HTA	TAFEI
Repertory grids	Heuristics	Observation

The methods are also more or less appropriate for different stages of the design process. Concept refers to very early, before any blueprints have been defined. Design is when there is a formalised version of the product on paper. Prototype is if there is either a computer simulation or hard-built version of the product available, but it is not yet in the marketplace. Operational refers to the product's commissioning and maintenance.

Concept	Design	Prototype	Operational
Checklists	KLM	KLM	KLM
HTA	Link analysis	Link analysis	Link analysis
Repertory grids	Checklists	Checklists	Checklists
Interviews	PHEA	PHEA	PHEA
Heuristics	HTA	Observation	Observation
	Repertory grids	Questionnaires	Questionnaires
	TAFEI	HTA	HTA
	Layout analysis	Repertory grids	Repertory grids
	Interviews	TAFEI	TAFEI
	Heuristics	Layout analysis	Layout analysis
		Interviews	Interviews
		Heuristics	Heuristics

Now you should assess how much time you have available for analysis. This will be relative to the product you are assessing, so terminology may seem vague here. However, using the car radio as a rough guide, 'not much' is less than 2 hours; 'some' is 2–6 hours; and 'lots' is more than 6 hours. This excludes time for training and practice.

Not much time	Some time	Lots of time
Checklists	KLM	KLM
Observation	Link analysis	Link analysis
Questionnaires	Checklists	Checklists
Layout analysis	Observation	PHEA
Heuristics	Questionnaires	Observation
	Repertory grids	Questionnaires
	Layout analysis	HTA
	Interviews	Repertory grids
	Heuristics	TAFEI
		Layout analysis
		Interviews
		Heuristics

Some methods are better to use on potential end-users of the product, although this is not essential. A couple of methods need two analysts present to complete.

- Better on end-users:
  - observation
  - questionnaires
  - repertory grids
  - interviews
- Requires two analysts:
  - observation
  - interviews

Finally, you should choose your method according to the output you require: errors, performance times, usability or design.

Errors	Times	Usability	Design
PHEA	KLM	Checklists	Link
Observation	Observation	Questionnaires	Checklists
TAFEI		HTA	PHEA
		Repertory grids	Rep grids
		Interviews	TAFEI
		Heuristics	Layout
			Heuristics

The further through the design cycle you are, and the more time you have available, the greater your choice of methods. You will also probably be aware that a number of decision paths result in the exclusion of all methods. This is why the software version is easier to use, as it eliminates these paths online. There is an alternative way to choose which methods to use, and that is by the utility analysis equation described in Section 4. This is a cost-benefit analysis which can provide the relative merits in financial terms of one method over another; it may be worthy of consideration. There is much that can be done to improve existing practices in ergonomics. The selection of methods to evaluate designs should depend on five factors:

- Accuracy of methods
- Criteria to be evaluated (e.g. time, errors, movements)
- Acceptability and appropriateness of the methods
- Abilities of the designers involved in the process
- Cost-benefit analysis of methods

Establishing the validity of the methods makes good commercial sense, especially when one considers the costs associated with poor decisions, for example:

- Poor productivity of design personnel
- Knock-on effects of difficult devices in terms of consumer perceptions
- Wasted design effort
- Loss of sales
- Increased customer support
- Cost of redesign

We therefore argue that organisations need to pay careful attention to all the stages in the method selection process, from developing appropriate selection criteria, through choosing methods and making the design decision, to validating the process as a whole (Figure 1.4).

Method selection is a closed-loop process with two feedback loops. The first feedback loop informs the selectors about the adequacy of the methods to meet the demands of the criteria, and the second feedback loop provides feedback about the adequacy of the device assessment process as a whole. The main stages in the process are identified as follows:

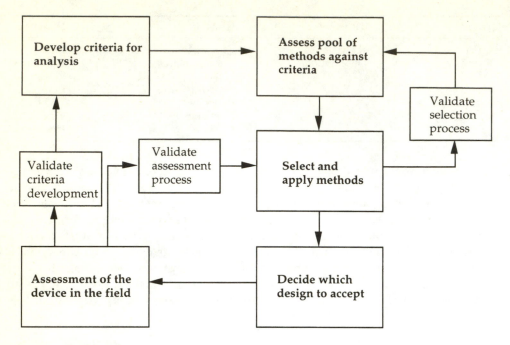

Figure 1.4 Validating the method selection process.

- *Determine criteria*: the criteria for assessment are identified.
- *Compare methods against criteria*: the pool of methods are compared for their suitability.
- *Apply methods*: the methods are applied to the assessment of a device.
- *Make decisions*: the resultant data are evaluated and the device design is chosen.
- *Validate*: the procedures for developing and assessing the criteria are validated.

Assessment criteria will be developed according to what information is required (e.g. speed of performance, error potential, user satisfaction and general aspects of device usability). Assessment methods may be determined by factors such as time available, access to end-users and cost of using the method.

Mini bibliography

Norman, D. A. (1988) *The Psychology of Everyday Things*, New York: Basic Books.

Section 2

CASE STUDIES

1. KEYSTROKE LEVEL MODEL (KLM)

Overview

KLM (Card, Moran and, Newell, 1983) is a simple technique which attempts to predict the time to execute a task given error-free performance, and with known time parameters of the system and the user. It was originally devised for human–computer interaction (HCI), so much of the language is tailored to computing.

There are four motor operators in KLM – keystroking, pointing, homing and drawing – one mental operator and one operator for system response. Each of these operators has an associated nominal time, derived by experiment (although drawing and response times are variable). It is thus a simple matter of determining the components of the task in question and summing the times of the associated operators to arrive at an overall task time prediction.

The physical operators are usually determined first (by observation or a similar method), and then a set of heuristic rules is used to place the mental operators. The pointing and drawing operators have little use outside HCI, so they have not been included in this book.

Although KLM predictions correlate very well with observed performance times on a car radio, empirical evidence suggests there is also a significant difference between predicted and observed. This difference probably represents extra time associated with driving, and would be constant across all in-car devices. The strength of KLM therefore lies in choosing between alternative designs for an interface on the basis of performance times. As this would be a relativistic judgement, the difference between predicted and observed is nullified. If, however, the analyst desires absolute predictions of performance times, it is recommended that a multiplier is used to bring the predictions more in line with what actually happens. Future research is needed to confirm these assumptions.

As KLM is applied to actual tasks, a concrete design is needed before the analysis can be carried out.

Procedure

Total execution times for specified operations are given as the sum of standard action times involved, e.g.

$$T_{exec} = T_m + T_h + T_k + T_r$$
$$= 1.35 + 0.4 + 0.2 + T_r$$
$$= 1.95s + T_r$$

For increased accuracy in predictions with in-car devices, multiply this total by 1.755, so

$$T_{exec} = 1.755(1.95 + T_r)$$

Heuristic rules for placing the mental (M) operations

Begin with a method of encoding that includes all physical operations and response operations. Use rule 0 to place candidate Ms, and then cycle through rules 1 to 4 for each M to see whether it should be deleted.

- Rule 0: insert Ms in front of all Ks that are not part of argument strings proper (e.g. text or numbers).
- Rule 1: if an operator following an M is *fully anticipated* in an operator just previous to the M then delete the M.
- Rule 2: if a string of MKs *belongs to a cognitive unit* (e.g. the name of a command) then delete all Ms but the first.
- Rule 3: if a K is a *redundant terminator* (e.g. the terminator of a command immediately following the terminator of its argument) then delete the M in front of it.
- Rule 4: if a K *terminates a constant string* (e.g. a command name), then delete the M in front of it; but if the K terminates a variable string (e.g. an argument string) then keep the M in front of it.

KLM Standard Performance Times

Action	Definition	Label	Standard time (s)
Button press	Pressing a button	K	0.2
Homing	Positioning finger over the button	H	1.1
Mental operation	Performing cognitive processes before pressing the button	M	1.35
System response time	Performance characteristics of the system	R	k

Source: Adapted from Card, Moran and Newell (1983).

Advice

KLM is used for predicting error-free performance times for defined tasks. Therefore, an exhaustive list of tasks with the product under analysis must be made. If the goal is to compare two or more different designs, care must be taken to ensure the tasks can be performed on all of the devices.

To determine performance times, one may observe error-free performance with the system, or a simpler walk-through will suffice in most cases (perhaps using the product manual).

For each task, performance times are calculated by determining the component operations involved. First insert the physical operations. It is unlikely with in-car devices that drawing or pointing operations will be used; most will simply involve homing and/or button presses. A simple 'one-button' task will therefore involve one homing action and one keystroke.

Next insert appropriate system response times. These are variable, so manufacturer's specifications will be needed if accuracy is paramount; otherwise, an expert estimation will suffice.

Finally, insert the mental operations. This is thinking time, and there is a set of heuristic rules for placing these operations. For most simple in-car tasks, though, it is likely that only one mental operation will be involved in each task.

Once all the operations have been inserted, and system response times noted, a simple addition reveals the total time for error-free performance of that task. If you are comparing devices or designs, it is sufficient to stop at this point and determine which is the best (i.e. the one with the quickest performance times). If you want a more accurate prediction of absolute performance times, it is recommended that the final sum is multiplied by 1.755 to approximate actual performance time.

As the KLM standard performance times were based on HCI operations, there may be occasions with in-car devices where problems may arise. For instance, there is no operator for 'turning a knob'. In such cases it is up to the discretion of the analyst to decide how to deal with it. In the instance of turning a knob, inserting a single key-stroking operator seems to be valid. Future research will determine more effective ways of coping with these situations.

Advantages and disadvantages

✓ Very straightforward and quick to apply.
✓ Little training required.
✓ Seems suited to motorway driving.
✗ Limited predictions.
✗ Restrictive – tailored to HCI.
✗ Needs validation outside HCI.

Mini bibliography

Baber, C., Hoyes, T. and Stanton, N. A. (1993) 'Comparison of GUIs and CUIs: appropriate ranges of actions and ease of use', *Displays*, **14**, 4, pp. 207–15.

Card, S. K., Moran, T. P. and Newell, A. (1983) *The Psychology of Human–Computer Interaction*, Hillsdale, NJ: Lawrence Erlbaum.

Eberts, R. (1997) Cognitive modeling, in Salvendy, G. (ed.) *Handbook of Human Factors and Ergonomics*, 2nd edn, New York: John Wiley, pp. 1328–74.

Jordan, P. W (1998) *An Introduction to Usability*, London: Taylor & Francis.

Keystroke Level Model (KLM): pros and cons

Reliability/validity ✳✳✳

By far the best technique statistically, KLM performs outstandingly on all three measures of reliability and validity. Although predicted task times correlate very well with those actually observed, there is also a significant difference between these data, indicating a constant 'thinking time' to be allowed for with the extra driving task.

Resources ✱✱

KLM is a moderately time-consuming method, taking longer to practice than to train. Execution times dramatically improve on the second application. A prepared HTA is advantageous in carrying out this analysis.

Usability ✱✱✱

KLM performed very well on the usability ratings, particularly for resource usage. Once the technique is learned, it is quite easily applied.

Efficacy

As this is a task-based analysis, it is necessary to have a formal design, although a prototype is not needed. Performance times are the primary output, so this is a good technique for comparing different potential designs. If specific performance times are required, it is recommended that a constant is added to allow for simultaneous tasks (e.g. driving).

Example

Task	Execution time (s)		$t_1 - t_2$ (s)
	t_1 for design 1	t_2 for design 2	
Switch on	MHKR=2.65+1=3.65	MHKR=2.65+1=3.65	0
Adjust volume	MHKR=2.65+0.1=2.75	MHKR=2.65+0=2.65	+0.1
Adjust bass	MHKHKR=3.95+0.2=4.15	MHKR=2.65+0=2.65	+1.5
Adjust treble	MHKKHKR=4.15+0.3=4.45	MHKR=2.65+0=2.65	+1.8
Adjust balance	MHKKHKR=4.15+0.3=4.45	MHKKR=2.85+0.1=2.95	+1.5
Choose new preset	MHKR=2.65+0.2=2.85	MHKR=2.65+0.2=2.85	0
Use seek	MHKR=2.65+1=3.65	MHKR=2.65+1=3.65	0
Use manual search	MHKHKR=3.95+1=4.95	MHKR=2.65+1=3.65	1.3
Store station	MHKR=2.65+1=3.65	MHKR=2.65+3=5.65	−2
Insert cassette	MHKR=2.65+1=3.65	MHKR=2.65+1=3.65	0
Autoreverse and FF	MHKRHKRKR=4.15+5=9.15	MHKRKRKR=3.05+5=8.05	1.1
Eject cassette	MHKR=2.65+0.5=3.15	MHKR=2.65+0.3=2.95	0.2
Switch off	MHKR=2.65+0.5=3.15	MHKR=2.65+0.7=3.35	−0.2
Total	53.65	48.35	5.3

This table represents the calculations for execution times of standard tasks across the two different radio designs. Design 1 is the Ford 7000 RDS EON, design 2 is the Sharp RG-F832E. On this set of tasks the Ford design takes around 5 s longer to complete. Analysing the operators suggests this extra time is very much taken up by the moded nature of the device. System response times have been largely estimated in the table. Standard response times (e.g. for radio tuning) were applied to both designs.

KLM flowchart

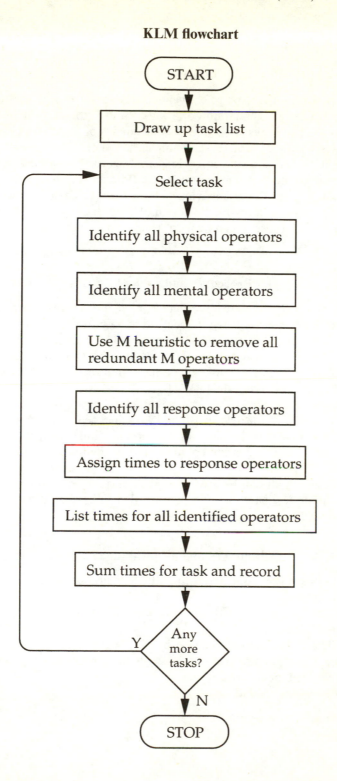

2. LINK ANALYSIS

Overview

Link analysis uses spatial diagrams to convey its message; it is specifically aimed at improving design. It relies on observation or a walk-through to establish links between components in the system. Links are defined as movements of attentional gaze or position between parts of the system, or communication with other system elements. Originally aimed at process control rooms, link analysis can be applied to individual devices by considering hand or eye movements, etc. Product design can most probably benefit from considering activity or behavioural links between device elements.

Link analysis is a way of improving the design of a product by directly understanding processes involved in its use. It can be used when formal specifications about product design are available.

Procedure and advice

Link analysis can be used to analyse either hand or eye movements. In the case of in-car devices, it is probably most sensible to focus on hand movements.

A representative set of tasks with the device should be listed, then a simple task walk-through by the analyst will provide the necessary information. Data should be recorded on which components are linked by a hand movement in any task, and how many times this link occurs during the task set.

This output can be represented by either a link diagram or a link table. The link diagram is a schematic layout of the device, overlaid with lines representing the course and number of links. A link table is the same information in tabular format – components take positions at the heads of rows and columns, and the numbers of links are entered in the appropriate cells.

Both these representations can be used to suggest revised versions of the layout of components for the device, based on the premise that links should be minimised in length, particularly if they are important or frequently used.

Advantages and disadvantages

✓ Very straightforward technique, requiring little formal training.
✓ Few resources required.
✓ Immediately useful output, with the process of analysis actually generating improvements to design.
✗ Preliminary data collection required (e.g. by observation).
✗ Only considers basic physical relationships; cognitive processes and error mechanisms not accounted for.
✗ Output not easily quantifiable.

Mini bibliography

Drury, C. G. (1990) Methods for direct observation of performance, in Wilson, J. and Corlett, E. N. (eds) *Evaluation of Human Work: A Practical Ergonomics Methodology*, 2nd edn, London: Taylor & Francis, pp. 45–68.

Kirwan, B. (1994) *A Guide to Practical Human Reliability Assessment*, London: Taylor & Francis.

Kirwan, B. and Ainsworth, L. K. (eds) (1992) *A Guide to Task Analysis*, London: Taylor & Francis.

Sanders, M. S. and McCormick, E. J. (1993) *Human Factors in Engineering and Design*, 7th edn, New York: McGraw-Hill.

Sanderson, P. M. and Fisher, C. (1997) Exploratory sequential data analysis: qualitative and quantitative handling of continuous observational data, in Salvendy, G. (ed.) *Handbook of Human Factors and Ergonomics*, New York: John Wiley, pp. 1471–1513.

Link Analysis: pros and cons

Reliability/Validity ✱✱

Link analysis performed particularly well on measures of intra-rater reliability and predictive validity. Unfortunately, the inter-rater reliability of the method let it down somewhat, being less than perfect.

Resources ✱✱

This method was relatively fast to train and practice, and execution time was moderate in comparison to the other techniques. It is usually helpful to be in possession of an HTA for the device under analysis, but this is not essential.

Usability ✱✱

Link analysis received average ratings of consistency, although these improved at trial 2. Resource usage was rated slightly higher, again improving over time.

Efficacy

As link analysis essentially deals with the layout of an interface, it can be used as soon as a formalised design is available. It is particularly useful for rearranging the layout of components on a device, or for comparing alternative designs.

Examples

The analyses are abbreviated and based on a standard subset of tasks which are common to both designs:

1. Switch on
2. Adjust volume
3. Adjust bass

4. Adjust treble
5. Adjust balance
6. Choose new preset
7. Use seek, then store station
8. Use manual search, then store station
9. Insert cassette
10. Autoreverse, then fast forward
11. Eject cassette and switch off

Redesigns are offered on the basis of analyses. Very little is changed on the Ford radio, suggesting the original satisfied the principles of link analysis well. The Sharp radio has been subject to more redesign, indicating the layout could have been conceived better.

Ford radio: initial design

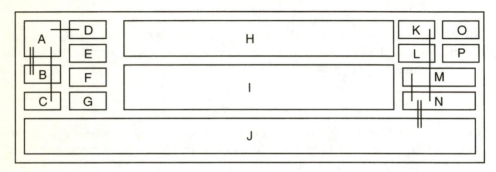

A - On/Off/Vol
B - Bass/Treb
C - Fade/Bal
D - Eject
E - Dolby
F - News
G - TA
H - Cassette Door

I - Display
J - Presets
K - Tape
L - PTY
M - Menu
N - Seek
O - CD
P - AM/FM

Ford radio: initial design cont'd

	A	B	C	D	E	F	G	H	I	J	K	L	M	N	O	P
A	x															
B	2	x														
C	1		x													
D	1			x												
E					x											
F						x										
G							x									
H								x								
I									x							
J										x						
K											x					
L												x				
M													x			
N										2	1		1	x		
O															x	
P																x
	A	B	C	D	E	F	G	H	I	J	K	L	M	N	O	P

Ford radio: revised design

Sharp radio: initial design

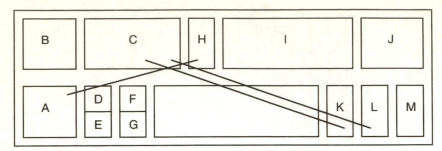

A = On/Off/Volume/Balance/Fader
B = Treble/Bass
C = Station Preset Buttons
D = FM Mono/Stereo Button
E = DX/Local Button
F = Band Selector Button
G = ASPM/Preset Memory Scan Button

H = Tape Eject Button
I = Cassette Compartment
J = Fast Wind/Programme Buttons
K = Tuning Up/Down Buttons
L = Tuning Scan/Seek Buttons
M = CD Input Socket

	A	B	C	D	E	F	G	H	I	J	K	L	M
A	x												
B		x											
C			x										
D				x									
E					x								
F						x							
G							x						
H	1							x					
I								x					
J									x				
K			1							x			
L			1								x		
M													x

Sharp radio: revised design

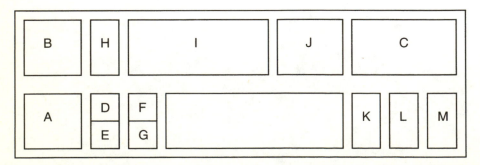

Link analysis flowchart

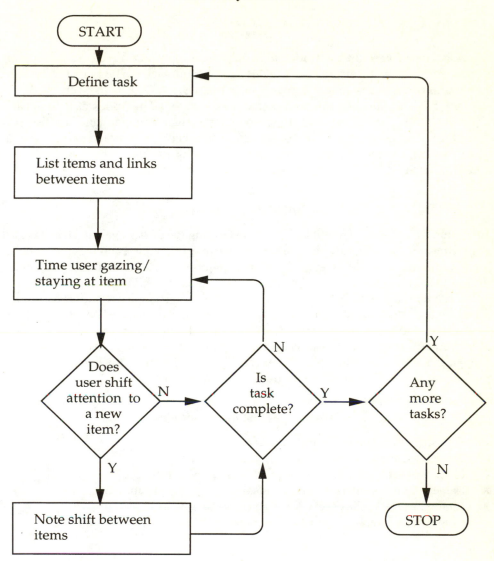

3. CHECKLISTS

Overview

Checklists need very little introduction; they are simply predefined lists of points against which an assessor can check the design of a product. There must therefore be some form of the device (either on paper or in prototype) available to be checked.

Many checklists already exist, and examples may be found in Woodson, Tillman and Tillman (1992). The current project uses the human engineering design checklist. It is up to the individual analyst to choose an appropriate checklist for each particular situation. We suggest that for most automotive applications, the human engineering design checklist, or the Ravden and Johnson (1989) checklist will suffice.

Procedure and advice

Executing a checklist is a simple matter of inspecting the device against each checklist item. However, checklists are also malleable; they may be adapted and modified according to the demands of the analysis. This aspect requires a certain amount of skill on the part of the analyst, for it is probably the source of most variability in checklist analyses.

For example, only one section of the human engineering design checklist is relevant to assessing in-car devices, the section on console and panel design. Thus it is perfectly reasonable (and not at all arduous) to extract the relevant items from a checklist and effectively construct a customised checklist for the device under analysis. Similarly, even though the Ravden and Johnson (1989) checklist was designed for human–computer interaction, it is easily adapted to the usability of other devices.

Advantages and disadvantages

- ✓ Extremely easy and quick technique.
- ✓ Procedural analysis ensures all aspects are covered.
- ✓ Based on established knowledge about human performance.
- ✗ Largely anthropometric – errors and cognitive problems not handled.
- ✗ Limited transferability – excessive generality/specificity.
- ✗ Context and interactions of tasks not considered.

Mini bibliography

Butters, L. M. (1998) Consumer product evaluation: which method is best? A guide to human factors at Consumers' Association, in Stanton, N. A. (ed.) *Human Factors in Consumer Products*, London: Taylor & Francis, pp. 159–71.

Dul, J. and Weerdmeester, B. (1993) *Ergonomics for Beginners: A Quick Reference Guide*, London: Taylor & Francis.

Johnson, G. I. (1996) The usability checklist approach revisited, in Jordan, P. W., Thomas, B., Weerdmeester, B. A. and McClelland, I. L. (eds) *Usability Evaluation in Industry*, London: Taylor & Francis, pp. 179–88.

Kirwan, B. and Ainsworth, L. K. (eds) (1992) *A Guide to Task Analysis*, London: Taylor & Francis.

Majoros, A. E. and Boyle, E. (1997) Maintainability, in Salvendy, G. (ed.) *Handbook of Human Factors and Ergonomics*, 2nd edn, New York: John Wiley, pp 1569–92.

24

Ravden, S. J. and Johnson, G. I. (1989) *Evaluating Usability of Human–Computer Interfaces: A Practical Method*, Chichester: Ellis Horwood.

Sinclair, M. A. (1990) Subjective assessment, in Wilson, J. R. and Corlett, E. N. (eds) *Evaluation of Human Work: A Practical Ergonomics Methodology*, 2nd edn, London: Taylor & Francis, pp. 69–100.

Woodson, W. E., Tillman, B. and Tillman, P. (1992) *Human Factors Design Handbook*, 2nd edn, New York: McGraw-Hill.

Checklists: pros and cons

Reliability/Validity ✱

Although checklists perform quite poorly on intra-rater reliability, inter-rater reliability surpasses all methods, and overall predictive validity is also fairly good. This is probably largely due to the highly structured nature of checklists.

Resources ✱✱✱

Checklists are one of the quickest techniques to train, practice and apply. Very little explanation is needed, and execution is a simple matter of ticking boxes.

Usability ✱✱✱

Checklists were regarded as the most consistent of all the methods used here. Resource usage, an indicator of ease-of-use, was also rated well.

Efficacy

Depending on which checklists are used (the literature offers many), they can theoretically be applied at any point during design. In the early stages, they can essentially be used as guidelines. But the checklist used in this study can only be applied realistically from the prototyping stage onwards.

Examples

Here are selected items from the human engineering design checklist (Woodson, Tillman and Tillman, 1992) which are relevant and/or marginal (unsatisfactory) for the car radios under analysis.

Ford radio

4. Console and panel design
4.1 Displays
4.1.2 Labeling

    a. Trade names and other irrelevant information not deleted.

    j. Some labels only relevant in one mode, so do not clearly indicate the function being controlled.

    l. PTY abbreviation is not common or meaningful.

    m. Dolby symbol too abstract to be used as label.

    n. Lettering on panels is not standard black, rather white on black.

4.1.3 Coding

    t. Instrument panels are dark grey rather than medium grey.

4.1.6 Indicator and legend lights

    l. Functional groups identified by recesses rather than black lines.

    o. Illumination of preset buttons is not bright enough.

    p. No dimming control is provided for transilluminated indicators.

    w. Most button surfaces are convex rather than concave.

4.1.8 Levers

    m. On/Off/Volume knob too small according to checklist dimensions.

4.2 Control/Display Relationships
4.2.1 Arrangements

    c. Display shows volume, but this is to the side of control knob rather than above it.

Sharp radio

4. Console and panel design
4.1 Displays
4.1.1 Principles

    e. Crucial visual checks not identified by attention-getting devices (e.g. visual or aural signals).

    g. Probability of confusion among instruments.

4.1.2 Labeling

    a. Trade names and other irrelevant information not deleted.

    b. Not easy to read under some expected conditions of illumination.

4.1.4 Scales, dials, counters

    a. Numbers and letters too small for accurate reading at normal distance.

    b. Reflected light may create illusion warning is ON or obscure reading.

4.1.6 Indicator and legend lights

> k. Displays not arranged in relation to one another to reflect the sequence of use or the functional relations of the components they represent, in that order of preference.
>
> l. Distinct, functional areas not set apart for purposes of ready identification.
>
> p. No dimming control for transilluminated indicators.
>
> t. No provision for bulb removal from the front of the display panel without the use of tools, or other equally rapid and convenient means.
>
> w. Button surfaces are flat rather than concave to fit the finger.
>
> x. Buttons do not provide sufficient tactile feedback to indicate that the control has been activated.
>
> y. No cover guard is provided to prevent of accidental activation.
>
> bb. Button displacement is less than one-eighth inch.
>
> cc. Button resistance is less than 10 ounces.

4.1.8 Levers

> m. For knobs grasped by the fingertips:
>> 1) Minimum depth less than three-quarter inch
>> 2) Minimum diameter less than one inch

4.2 Control/Display Relationships

4.2.1 Arrangements

> a. Controls having sequential relations, or having to do with a particular function or operation, or which are operated together, are not grouped together or with the associated displays.
>
> b. Controls associated with transilluminated indicators, are not located so as to be immediately and unambiguously associated with the indicator.
>
> c. Controls associated with transilluminated indicators are not located below the indicator.
>
> d. Location of some controls may cause obscuration of the display by the hand normally used for setting the control.

4.2.2 Precautions

> a. Some controls are located so that the operator may hit or move them accidentally in the normal sequence of control movements.
>
> d. No interlocks are provided so that extra movement of the prior operation of a related or locking control is required.
>
> e. No resistance is built into the controls so that definite or sustained effort is required to actuate them.

Checklists flowchart

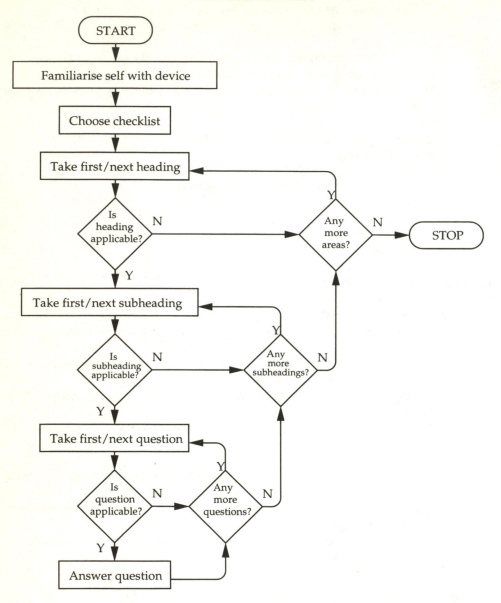

4. PREDICTIVE HUMAN ERROR ANALYSIS (PHEA)

Overview

PHEA 9 (also called SHERPA) is a development of HTA in that it uses each bottom-level task of the hierarchy as its inputs. These tasks are categorised according to a pre-determined taxonomy and form the basis of subsequent error identification. Thus the first step of a PHEA must be to devise an HTA if one is not already available.

A human error taxonomy is used to classify tasks into one of five error types (action, retrieval, checking, selection, information communication). The analyst then refers to the taxonomy to assess credible error modes for each task.

For each potential error, the analyst then evaluates consequentiality, ordinal probability and criticality. Then based on the subjective judgement of the analyst, possible remedial actions are proposed, along with recovery steps at which they may be effected.

This process occurs for each bottom-level task of the HTA, and the entire procedure is illustrated by means of a flowchart.

Procedure and advice

For every bottom-level task in the HTA (i.e. every operation), the following procedure is adopted.

First, assign it to one of the classes of error provided in the PHEA taxonomy. A well-designed HTA should make it obvious which class each operation belongs to. Let us suppose the operation we are looking at is an 'action'.

Each class has associated with it a number of error types which may or may not occur in a given context. It is the job of the analyst to decide whether any of the types are credible for the current situation. The 'action' error class has nine error types, so for our example operation we need to determine which of them are credible. If an error type is not credible, it is excluded from further analysis. However, if it is credible we carry on.

For all credible error types associated with an operation, the analyst should note down a description of the error, any associated consequences, whether it is recoverable (and if so, at which task step), the ordinal probability of it occurring, its criticality, and any proposed remedies. This information can be tabulated for ease of documentation. Ordinal probability (P) of an error can be categorised as low (hardly ever occurs), medium (has occurred once or twice) or high (occurs fairly frequently). Criticality (C) is all or none; it can be defined by the analyst, bearing in mind the system under consideration.

This procedure is repeated for every bottom-level task in the HTA.

Advantages and disadvantages

✓ Structured and comprehensive procedure, yet maintains usability.
✓ Taxonomy prompts analyst for potential errors.
✓ Encouraging validity and reliability data.
✓ Substantial time economy compared to observation.
✓ Error reduction strategies offered as part of the analysis, in addition to predicted errors.

✗ Can be tedious and time-consuming for complex tasks.

✗ Extra work involved if HTA not already available.

✗ Does not model cognitive components of error mechanisms.

✗ Some predicted errors and remedies are unlikely or lack credibility, thus posing a false economy.

✗ Current taxonomy lacks generalisability.

Mini bibliography

Baber, C. and Stanton, N. A. (1996a) 'Human error identification techniques applied to public technology: predictions compared with observed use', *Applied Ergonomics*, **27**, 2, pp. 119–31.

Embrey, D. E. (1993) 'Quantitative and qualitative prediction of human error in safety assessments', *Institute of Chemical Engineers Symposium Series*, **130**, pp. 329–50.

Kirwan, B. (1990) Human reliability assessment, in Wilson, J. R. and Corlett, E. N. (eds) *Evaluation of Human Work: A Practical Ergonomics Methodology*, 2nd edn, London: Taylor & Francis, pp. 921–68.

Kirwan, B. (1994) *A Guide to Practical Human Reliability Assessment*, London: Taylor & Francis.

Stanton, N. A. (1995) 'Analysing worker activity: a new approach to risk assessment?' *Health and Safety Bulletin*, **240**, Dec., pp. 9–11.

Stanton, N. A. and Stevenage, S. V. (1998) 'Learning to predict human error: issues of acceptability, reliability and validity', *Ergonomics*, **41**, 11, pp. 1737–56.

Predictive Human Error Analysis (PHEA): pros and cons

Reliability/Validity ✶✶

PHEA performed well in terms of predictive validity, although intra-rater reliability failed to reach significance. Inter-rater reliability was quite good on the second trial. It would be dubious to draw conclusions about the validity of this technique without supporting reliability data.

Resources ✶

One of the longest methods to train and practice in – analysts' rehearsal times far outweigh their actual tutoring time. In terms of execution times, PHEA is again very high, although this improves somewhat on trial 2. It must be taken into account that PHEA is based upon HTA, so unless this analysis has already been carried out, the resources invested in HTA must be added to those mentioned here.

Usability ✶✶

PHEA was rated as highly consistent by our trainees. Resource usage was considered worse than average on the first application, however impressions were improved by trial 2.

Efficacy

As PHEA is based on HTA (see section 7), any restrictions applied to that technique also apply here. Therefore, PHEA will only be applicable when there is a design available to work with. It is most useful either before or around the prototyping stage, as the output (predicted errors) may be used in redesign.

Examples

Ford radio

Task step	Error mode	Description	Consequence	Recovery	P	C	Remedies
1	C1	Fail to check whether unit is already on	Turn unit off instead of on	Immediate	L		Separate on/off buttons
2	A6	Press wrong button	Unit is not switched on	Immediate	L		Make on/off button more accessible
	A7	Turn on/off control instead of pressing it	Unit is not switched on	Immediate	L		Distinguish on/off from volume control
3.1	C1	Fail to check waveband display	Listening to tape instead of radio	3.3		M	Make radio display more prominent
3.2	A4	Press waveband selector too many times	Not listening to desired waveband	3.3		M	Have delay buffer on waveband selector
	A6	Press wrong button	Listening to tape instead of radio	3.3		M	Make waveband selector more distinct
	A8	Fail to press waveband selector	Listening to tape instead of radio	3.3		M	Make radio display more prominent
3.3	C1	Fail to check station	Not listening to desired station	Immediate	L		Make radio display more prominent
3.4.1	C1	Fail to check wavelength	Not listening to desired wavelength	3.4.2	L		Make radio display more prominent
3.4.2	S1	Fail to select wavelength	Not listening to desired wavelength	3.4.3	L		Have prompt for user to select wavelength
	S2	Select wrong wavelength	Not listening to desired wavelength	3.4.3	L		Make radio display more prominent
3.4.3.1	A1	Press seek for too long	Miss desired station	Immediate	M		Have one-touch operation
	A6	Press wrong button	Fail to tune radio	Immediate	L		Make seek button more distinct
	A8	Fail to press seek button	Fail to tune radio	3.4.3.2	L		Have prompt for user to tune radio
3.4.3.2	A1	Do not press preset button long enough	Fail to store station	Immediate	H		Decrease threshold time
	A6	Press wrong button	Fail to store station	Immediate	L		Make preset buttons more distinct
	A8	Fail to press preset button	Fail to store station	Immediate	L		Have prompt for user to store station
3.4.4	A1	Do not press button long enough	Autostore fails	Immediate	H		Decrease threshold time
	A6	Press wrong button	Autostore fails	Immediate	L		Make autostore button more distinct
	A8	Fail to press autostore button	Autostore fails	Immediate	L		Have prompt for user to press autostore button

Ford radio cont'd

Task step	Error mode	Description	Consequence	Recovery	P	C	Remedies
3.4.5.1	A1	Press menu button for too long	Radio enters secondary menu	Immediate		L	Have separate operation for secondary menu
	A4	Press menu too many times	Not in manual tuning mode	Immediate		M	Have 'back' button on menu
	A6	Press wrong button	Not in manual tuning mode	Immediate		L	Make menu button more distinct
	A8	Fail to press menu button	Not in manual tuning mode	Immediate		M	Have prompt for user to press menu button
3.4.5.2	A1	Press seek button for too long	Miss desired station	Immediate		M	Decrease speed of seek
	A2	Wait too long before pressing seek	Mode reverts to radio	Immediate		M	Have longer delay on mode selection
	A6	Press wrong button	Fail to tune radio	Immediate		L	Make seek buttons more distinct
	A8	Fail to press seek button	Mode reverts to radio	Immediate		L	Have prompt for user to press seek button
3.4.5.3	A1	Do not press button long enough	Fail to store station	Immediate		H	Decrease threshold time
	A6	Press wrong button	Station not stored	Immediate		L	Make preset buttons more distinct
	A8	Fail to press preset button	Station not stored	Immediate		L	Have prompt for user to press preset button
3.4.6	A1	Press preset button for too long	Station stored inappropriately	Immediate		M	Confirm before storing station
3.5.1.1	A1	Press TA button for too long	Radio enters TA volume mode	Immediate		L	Have separate operation for TA volume
	A6	Press wrong button	Radio not in TA mode	Immediate		L	Make TA button more distinct
	A8	Fail to press TA button	Radio not in TA mode	Immediate		L	Have prompt for user to press TA button
3.5.1.2	C1	Fail to check volume setting	TA volume inappropriately adjusted	3.5.1.3		L	TA volume mode automatically invoked
	C3	Check radio volume instead of TA volume	TA volume inappropriately adjusted	3.5.1.3		M	TA volume mode automatically invoked
3.5.1.3.1	A1	Do not press TA button long enough	Not in TA volume mode	3.5.1.3.2		M	TA volume mode automatically invoked
	A6	Press wrong button	Not in TA volume mode	3.5.1.3.2		L	Make TA button more distinct
	A8	Fail to press TA button	Not in TA volume mode	3.5.1.3.2		L	Have prompt for user to press TA button
3.5.1.3.2	A2	Do not adjust volume quickly enough	Radio reverts to normal mode	Immediate		M	Stay in TA volume mode as default
	A8	Fail to adjust volume	Radio reverts to normal mode	Immediate		L	Stay in TA volume mode as default
3.5.2.1	A6	Press wrong button	Not in PTY mode	Immediate		L	Make PTY button more distinct
	A8	Fail to press PTY button	Not in PTY mode	3.5.2.2		L	Have prompt for user to press PTY button
3.5.2.2	C1	Fail to check display	Undesired station selected	3.5.2.3		L	Make radio display more prominent
	C5	Do not check display quickly enough	Radio reverts to normal mode	3.5.2.3		L	Stay in PTY mode as default

Ford radio cont'd

Task step	Error mode	Description	Consequence	Recovery	P	C	Remedies
3.5.2.3.1	A1	Press seek button for too long	Miss desired station	Immediate		M	Have one-touch operation
	A2	Do not press seek button quickly enough	Radio reverts to normal mode	Immediate		M	Stay in PTY mode as default
	A6	Press wrong button	Fail to tune radio	Immediate		L	Make seek button more distinct
	A8	Fail to press seek button	Fail to tune radio	Immediate		L	Have prompt for user to press seek button
3.5.2.3.2	C1	Fail to check display	Undesired station selected	3.5.2.4		L	Make radio display more prominent
	C5	Do not check display quickly enough	Radio reverts to normal mode	3.5.2.4		M	Stay in PTY mode as default
3.5.2.4	C1	Fail to check display	Undesired PTY selected	3.5.2.5		L	Make radio display more prominent
	C5	Do not check display quickly enough	Radio reverts to normal mode	3.5.2.5		M	Stay in PTY mode as default
3.5.2.5.1	A2	Do not adjust volume control quickly enough	Radio reverts to normal mode	3.5.2.5.2		M	Stay in PTY mode as default
	A8	Fail to adjust volume control	Undesired PTY selected	3.5.2.5.2		L	Have prompt for user to adjust volume control
	A10	Attempt other means to select PTY	Fail to select PTY	Immediate		H	Have prompt for user to adjust volume control
3.5.2.5.2	C1	Fail to check display	Undesired PTY selected	Immediate		L	Make radio display more prominent
	C5	Do not check display quickly enough	Radio reverts to normal mode	Immediate		M	Stay in PTY mode as default
3.5.3	A6	Press wrong button	Not in news mode	Immediate		L	Make news button more distinct
	A8	Fail to press news button	Not in news mode	Immediate		L	Have prompt for user to press news button
4.1.1	C1	Fail to check cassette deck	Do not know whether cassette is already in	4.1.2		L	Have prominent cassette deck indicator
4.1.2	A5	Miss cassette entry door	Cassette unsuccessfully inserted	Immediate		L	Make cassette door more accessible
4.1.3	A6	Press wrong button	Not in tape mode	Immediate		L	Make tape button more distinct
	A8	Fail to press tape button	Not in tape mode	Immediate		L	Have prompt for user to press tape button
4.2.1	C1	Fail to check tape position	Tape in wrong place	4.2.3		L	Introduce tape counter
4.2.2	A6	Press wrong button	Tape on wrong side	4.2.3		L	Make tape button more distinct
	A8	Fail to press tape button	Tape on wrong side	4.2.3		L	Make tape direction display more prominent
4.2.3	A1	Press seek buttons too long/not long enough	Tape in wrong place	Immediate		M	Introduce tape counter
	A6	Press wrong button	Tape in wrong place	Immediate		L	Make seek buttons more distinct
	A8	Fail to press seek buttons	Tape in wrong place	Immediate		L	Have prompt for user to press seek buttons

Ford radio cont'd

Task step	Error mode	Description	Consequence	Recovery	P	C	Remedies
4.3.1	A1	Press menu button for too long	Radio enters secondary menu	Immediate		L	Have separate operation for secondary menu
	A4	Press menu button too many times	Not in AMS mode	Immediate		M	Have 'back' button on menu
	A6	Press wrong button	Not in AMS mode	Immediate		L	Make menu button more distinct
	A8	Fail to press menu button	Not in AMS mode	Immediate		L	Have prompt for user to press menu button
4.3.2	A1	Press seek button for too long	Miss desired point on tape	Immediate		M	Have one-touch operation
	A2	Do not press seek button quickly enough	Radio reverts to normal mode	Immediate		M	Stay in AMS mode as default
	A6	Press wrong button	Tape not searching	Immediate		L	Make seek buttons more distinct
	A8	Fail to press seek button	Tape not searching	Immediate		L	Have prompt for user to press seek button
4.4	A6	Press wrong button	Tape not paused	Immediate		L	Make pause button more distinct
4.5	A6	Press wrong button	Dolby not selected	Immediate		L	Make Dolby button more distinct
4.6	A6	Press wrong button	Tape not ejected	Immediate		L	Make eject button more distinct
5.1.2.1	A1	Press menu button for too long	Radio enters secondary menu	Immediate		L	Have separate operation for secondary menu
	A4	Press menu button too many/not enough times	Not in AVC mode	Immediate		M	Have clearer menu structure
	A6	Press wrong button	Not in AVC mode	Immediate		L	Make menu button more distinct
5.1.2.2	A2	Do not press seek buttons quickly enough	Radio reverts to normal mode	Immediate		M	Stay in AVC mode as default
	A6	Press wrong button	AVC not adjusted	Immediate		L	Make seek buttons more distinct
	A8	Fail to press seek buttons	AVC not adjusted	Immediate		L	Have prompt for user to press seek buttons
5.1.2.3	A1	Press menu button for too long	Radio enters secondary menu	Immediate		L	Have separate operation for secondary menu
	A2	Do not press menu button quickly enough	Radio reverts to normal mode	Immediate		M	Stay in AVC mode as default
	A4	Press menu button too many/not enough times	Radio not in normal mode	Immediate		M	Have 'back'. button on menu
	A6	Press wrong button	Radio not in normal mode	Immediate		L	Make menu button more distinct
5.2.1	A4	Press bass button too many times	Not in bass mode	Immediate		M	Have separate bass button
	A6	Press wrong button	Not in bass mode	Immediate		L	Make bass button more distinct

Ford radio cont'd

Task step	Error mode	Description	Consequence	Recovery	P	C	Remedies
5.2.2	A2	Do not adjust volume control quickly enough	Radio reverts to normal mode	Immediate		M	Stay in bass mode as default
	A8	Fail to adjust volume control	Bass not adjusted	Immediate		L	Have prompt for user to adjust volume control
	A10	Attempt other means to adjust bass	Bass not adjusted	Immediate		H	Have prompt for user to adjust volume control
5.3.1	A4	Press bass button not enough times	Not in treble mode	Immediate		M	Have separate treble button
	A6	Press wrong button	Not in treble mode	Immediate		L	Make bass button more distinct
5.3.2	A2	Do not adjust volume control quickly enough	Radio reverts to normal mode	Immediate		M	Stay in treble mode as default
	A8	Fail to adjust volume control	Treble not adjusted	Immediate		L	Have prompt for user to adjust volume control
	A10	Attempt other means to adjust treble	Treble not adjusted	Immediate		H	Have prompt for user to adjust volume control
5.4.1	A4	Press fade button too many times	Not in fade mode	Immediate		M	Have separate fade button
	A6	Press wrong button	Not in fade mode	Immediate		L	Make fade button more distinct
5.4.2	A2	Do not adjust volume control quickly enough	Radio reverts to normal mode	Immediate		M	Stay in fade mode as default
	A8	Fail to adjust volume control	Fade not adjusted	Immediate		L	Have prompt for user to adjust volume control
	A10	Attempt other means to adjust fade	Fade not adjusted	Immediate		H	Have prompt for user to adjust volume control
5.5.1	A4	Press fade button not enough times	Not in balance mode	Immediate		M	Have separate balance button
	A6	Press wrong button	Not in balance mode	Immediate		L	Make fade button more distinct
	A2	Do not adjust volume control quickly enough	Radio reverts to normal mode	Immediate		M	Stay in balance mode as default
	A8	Fail to adjust volume control	Balance not adjusted	Immediate		L	Have prompt for user to adjust volume control
	A10	Attempt other means to adjust balance	Balance not adjusted	Immediate		H	Have prompt for user to adjust volume control

Sharp radio

Task step	Error mode	Description	Consequence	Recovery	P	C	Remedies
1	A4	Volume level adjusted inappropriately	Volume is at undesirable level	Immediate /4.2		M	Separate vol/on/off Preset start-up volume
	A7	Balance adjusted instead of on/off/vol	Unit is not switched on; balance settings altered	Immediate /4.3		H	Separate balance/on /off Lockout mechanism
2.1	C1	Omit check of station	Listening to wrong station	2.2		L	Untuned on start-up Reminder to check
	C2	Misidentify station	Listening to wrong station	2.2		M	RDS
2.2.1	S2	Select wrong preset	Listen to wrong station	2.2.2 on		H	Label buttons clearly Aide-memoire
2.2.2	A1	Preset button held too long	Undesired station storage	2.3		L	Confirm before storage
	A6	Press wrong button	Desired station not found	Immediate		H	Label buttons clearly Aide-memoire
2.3.1	C1	Omit check of wavelength	Unnecessary retuning	2.3.2		L	Display conspicuity (e.g. RDS)
	C3	Check wrong display	Wavelength not identified	2.3.2		L	Display conspicuity (e.g. RDS)
2.3.2	A2	Adjust wavelength late	Unnecessary retuning	2.3.4 on		L	Better manual/training
	A4	Overpressing button	Desired setting missed	Immediate		L	Responsive buttons Better manual/training
	A6	Press wrong button	Undesired action; desired wavelength omitted	Immediate		L	Label buttons clearly
	A8	Fail to change wavelength	Undesired wavelength; unnecessary retuning	2.3.4 on		L	Better manual/training
2.3.3	C1	Omit check of frequency	Unnecessary retuning	2.3.4		L	Display conspicuity (e.g. RDS)
	C3	Check wrong display	Frequency not identified	2.3.4		L	Display conspicuity (e.g. RDS)
2.3.4.1	A1	Autosearch too long	Miss desired frequency	Immediate		M	Better manual/training Alter search parameters
	A2	Adjust frequency before wavelength	Tuning undesired wavelength	Immediate /2.3.5		L	Confirmation prompt Better manual/training
	A6	Press wrong button	Undesired action; desired frequency omitted	Immediate		M	Label buttons clearly Clarify functions
	A8	Fail to adjust frequency	Undesired frequency; unnecessary retuning	2.3.5 on		L	Better manual/training
	A9	Incomplete search	Undesired frequency; unnecessary retuning	2.3.5 on		L	Better manual/training

Sharp radio cont'd

Task step	Error mode	Description	Consequence	Recovery	P	C	Remedies
2.3.4.2	A1	Press button too long	Switches to autosearch	Immediate	H		Alter button function Clarify function
	A2	Adjust frequency before wavelength	Tuning undesired wavelength	Immediate	L		Better manual/training Confirmation prompt
	A3	Tuning in wrong direction	Tuning inefficiently	Immediate	M		Increase compatibility Label buttons clearly Use knob instead
	A4	Overzealous use of button	Miss desired frequency	Immediate	M		Increase buffer
	A6	Press wrong button	Undesired action; desired frequency omitted	Immediate	M		Label button clearly Use knob instead
	A7	Use manual button for autosearch	Inefficient tuning	Immediate	M		Label button clearly Use knob instead
	A8	Fail to adjust frequency	Undesired frequency	2.3.5 on	L		Better manual/training
	A9	Incomplete frequency search	Undesired frequency; unnecessary retuning	2.3.5 on	L		Better manual/training
	A10	Autosearch on wrong button	Undesired action; desired omitted	Immediate	L		Label buttons clearly Use tuning knob
2.3.5	A1	Button press too short	Station not stored	Immediate	M		Better manual/training Decrease threshold time
	A2	Store station before frequency	Inappropriate storage	2.2 on	L		Better manual/training
	A6	Press wrong button	Station not stored /stored inappropriately	2.2 on	M		Label buttons clearly
	A7	Change station instead of storing	Station not stored; frequency lost	2.3 on	M		Better manual/training Confirmation prompt
	A8	Fail to store station	Station lost	2.2	L		Confirmation prompt Autostore prompt
	A10	Change to wrong station and not store	Undesired station; frequency lost	2.2	L		Confirmation prompt Autostore prompt
3.1	A5	Cassette insertion misaligned	Cassette will not fit	Immediate	L		Label entry door more clearly
3.2	C1	Omit check of cassette	Cassette in wrong position	3.3 on	M		Monitor/counter Clearer direction display
	C2	Incomplete check on cassette	Cassette in wrong position	3.3 on	M		Monitor/counter Clearer direction display
3.3	A10	Press FF/RWD buttons	Cassette in wrong position	3.4 on	M		Make autoreverse button distinct
3.4	A1	FF too long	Miss desired point	3.5	H		Monitor/counter
	A3	RWD instead of FF	Miss desired point	3.5	L		Label buttons clearly
	A6	Press wrong button	Miss desired point	3.5	L		Label buttons clearly
	A9	Incomplete search	Miss desired point	3.5	L		Monitor/counter
3.5	A1	RWD too long	Miss desired point	3.4	H		Monitor/counter
	A3	FF instead of RWD	Miss desired point	3.4	L		Label buttons clearly
	A6	Press wrong button	Miss desired point	3.4	L		Label buttons clearly
	A9	Incomplete search	Miss desired point	3.4	L		Monitor/counter

Sharp radio cont'd

Task step	Error mode	Description	Consequence	Recovery	P	C	Remedies
4.1	C1	Omit check of audio settings	Audio settings undesired	4.2 on	L		Preset preferences Mute on start-up
	C2	Incomplete check of audio settings	Audio settings undesired	4.2 on	L		Preset preferences
4.2	A3	Adjust volume wrong way	Undesired volume level	Immediate	L		Compatibility Discrete functions
	A4	Adjustment not enough/too much	Undesired volume level	Immediate	H		Scaled volume levels Preset preferences
	A6	Adjust balance/tone mistakenly	Undesired settings	4.3 on	M		Separate functions
	A7	Turn off instead of down	Unit off unintentionally	Immediate	L		Separate functions
4.3	A3	Adjust balance wrong way	Undesired balance level	Immediate	L		Compatibility Discrete functions
	A4	Adjustment not enough/too much	Undesired balance level	Immediate	H		Scaled balance levels Preset preferences
	A6	Adjust volume/tone mistakenly	Undesired settings	4.2 on	M		Separate functions
4.4	A3	Adjust tone wrong way	Undesired tone level	Immediate	L		Compatibility Discrete functions
	A4	Adjustment not enough/too much	Undesired tone level	Immediate	H		Scaled tone levels Preset preferences
	A6	Adjust volume /balance mistakenly	Undesired settings	4.2 on	M		Separate functions
5	A3	Turn up instead of off	Volume too high; unit still on	Immediate	L		Compatibility Separate functions
	A4	Mute instead of off	Unit still on	None	M		Eliminate mute Separate functions Increase conspicuousness
	A7	Adjust volume instead of switching off	Undesired volume; unit still on	Immediate	L		Separate functions
	A9	Incomplete operation; switch not triggered	Unit still on	Immediate	L		Increase conspicuousness Separate functions Improve switch

PHEA flowchart

START

Analyse task
using HTA

Take a task step from
the bottom level of the
analysis

Classify the task step into a type
from the taxonomy: action,
checking, information, retrieval,
selection

Are any
of the error types
credible?

Are there
any more task
steps?

Y

N

N

N

STOP

Y

For each error type:
• Describe the error
• Determine the consequence
• Enter recovery step
• Enter ordinal probability
• Enter criticality
• Propose remedy

Are there
any more error
types?

N

Y

5. OBSERVATION

Overview

There are many and varied observational techniques which fall into three broad categories: direct, indirect and participant observation. However, the applications and limitations are similar in each of them, and each generally requires at least two people (the observer and the participant). Although not essential it is an advantage if the participant is an end-user of the system. A working example of the product needs to exist for observation to be viable.

Observation can be very useful for recording physical task sequences or interactions between workers. Indeed, Baber and Stanton (1996b) have already considered the potential of observation as a technique for usability evaluation, and they provide helpful guidelines for researchers.

Procedure and advice

The observational method begins with a scenario – the observer should present the participant with the device and a list of tasks to perform. The observer may then sit back and record aspects of human–device interaction which are of interest. Typical measures are execution times and any errors observed. This information can be integrated into the design process for the next generation of devices.

Before commencing, the wise observer will draw up an observation sheet for use in data collection. Filling in cells on a table is quicker and easier than writing notes while your participant is performing a task.

Video observation can be a valuable tool, particularly with the computer-assisted analysis techniques now available. These computer analyses can greatly reduce the amount of data and the time to collect it.

One of the main concerns with observation is the intrusiveness of the observational method; it is well known that the behaviour of people can change purely as a result of being watched. People do get used to observers (or observational equipment) over time.

Another major problem is that one cannot infer causality from simple observation. That is, the data recorded must be a purely objective record of what actually happened, without any conjecture as to why.

Advantages and disadvantages

✓ Provides objective information which can be compared and ratified by other means.
✓ Can be used to identify individual differences in task performance.
✓ Gives 'real-life' insight into human–machine interaction.
✗ Very resource-intensive, particularly during analysis.
✗ Effect on observed party.
✗ Lab versus field trade-off (i.e. control versus ecological validity).
✗ Does not reveal any cognitive information.

Mini bibliography

Baber, C. and Stanton, N. A. (1996b) Observation as a technique for usability evaluations, in Jordan, P. W., Thomas, B., Weerdmeester, B. A. and McClelland, I. L. (eds) *Usability in Industry*, London: Taylor & Francis, pp. 85–94.

Drury, C. G. (1990) Methods for direct observation of performance, in Wilson, J. and Corlett, E. N. (eds) *Evaluation of Human Work: A Practical Ergonomics Methodology*, 2nd edn, London: Taylor & Francis, pp. 45–68.

Kirwan, B. and Ainsworth, L. K. (eds) (1992) *A Guide to Task Analysis*, London: Taylor & Francis.

Oborne, D. J. (1982) *Ergonomics at Work*, Chichester: John Wiley.

Observation – pros and cons

Reliability/Validity ✱✱

Observers' predictions of errors were consistent over time in the present study, although different raters were not consistent with each other. Predictive validity of the error scores was disappointing. Predictions of task times, on the other hand, correlated well with actual times, although there was also a significant difference between these data. This difference probably represents a constant 'thinking time', allowing for the fact that in the applied setting, participants have an extra task to deal with (i.e. driving).

Resources ✱✱✱

Thorough observational studies are usually very time-intensive. In this study, only one set of observations was used for the predictions, making for relatively low execution times. Training and practice times were also low for a technique which requires little explanation. It should be noted that at least two personnel would be required in executing this method.

Usability ✱✱

Observation received medium ratings of consistency by our trained analysts. Resource usage ratings were rather higher, indicating a preference for this method's ease of application.

Efficacy

To observe device use requires at least a prototype, and ideally observation is carried out in the marketplace with end-users. Output can then be fed back into the design process to refine future generations of the device.

Examples

Task list

1. Switch on
2. Adjust volume
3. Adjust bass
4. Adjust treble
5. Adjust balance
6. Choose a new preset station
7. Choose a new station using seek and store it
8. Choose a new station using manual search and store it
9. Insert cassette
10. Find the next track on the other side of cassette
11. Eject cassette
12. Switch off

Ford radio (30 participants)

Task	Errors observed	Freq 1	Freq 2	t_1 (s) mean (sd)	t_2 (s) mean (sd)[a]
1				4.64 (4.38)	4.05 (3.02)
2	Didn't turn knob enough to adjust	3	1	10.5 (7.67)	6.14 (1.88)
	Pressed seek	1			
3	Adjusted treble	1		20.4 (13.3)	10.1 (3.75)
	Adjusted volume	2			
	Pressed on/off	1			
4				15.7 (15.5)	8.55 (4.50)
5	Adjusted fade	2	5	18.1 (9.61)	11.9 (5.53)
	Adjusted bass	1			
	Didn't attempt – forgot how		1		
6	Used seek	1		7.14 (9.88)	3.86 (1.73)
7	Pressed preset and seek together	1	1	31.5 (24.7)	23.6 (10.7)
	Held seek button down	2	1		
	Interrupted seek by pressing preset		1		
	Pressed preset		1		
	Used manual tuning		1		
	Failed to store – didn't know how	10			
	Didn't hold preset long enough	10	4		
	Pressed seek to store	1			

Ford radio cont'd

Task	Errors observed	Freq 1	Freq 2	t_1 (s) mean (sd)	t_2 (s) mean (sd)[a]
8	Didn't know function	4		44.2 (23.8)	29.5 (13.6)
	Used seek	15	3		
	Held seek button down	2			
	Pressed preset	1			
	Failed to store – didn't know how	7			
	Didn't hold preset long enough	10	4		
	Hit two presets and storage failed		1		
9				3.64 (1.87)	3.18 (1.33)
10	Failed to stop FF/RWD	1	2	36.6 (19.2)	30.9 (16.1)
	Pressed seek instead of autoreverse	3			
	Pressed wrong direction	3	2		
	Turned tape over manually	6			
	Failed to seek	1	1		
11	Pressed twice		1	4.55 (2.77)	3.91 (1.69)
12				2.86 (1.21)	2.05 (0.576)

[a] sd = standard deviation

Sharp radio (2 participants)

Task	Errors observed	Freq 1	Freq 2	t_1 (s) mean (sd)	t_2 (s) mean (sd)[a]
1	Pressed knob instead of turning		1	5.37 (1.56)	5.27 (1.92)
2				4.66 (0.66)	4.67 (0.5)
3				10.03 (2.62)	6.77 (1.61)
4				9.64 (0.37)	4.39 (1.68)
5	Adjusted fade instead of balance	1	1	5.26 (0.11)	4.94 (0.8)
6				6.5 (5.78)	4.38 (0.45)
7	Pressed scan instead of seek	1		18.19 (3.86)	8.92 (0.32)
	Failed to store	1			
8	Failed to store	1	1	22.5 (9.06)	9.8 (1.63)
9				3.21 (0.66)	4.63 (1.06)
10	Didn't push reverse button hard enough	1		28.09 (10.54)	16.25 (3.86)
	Pressed fast forward only		1		
11				4.4 (0.71)	2.96 (0.64)
12				3.17 (0.68)	2.9 (0.03)

[a] sd = standard deviation

Observation flowchart

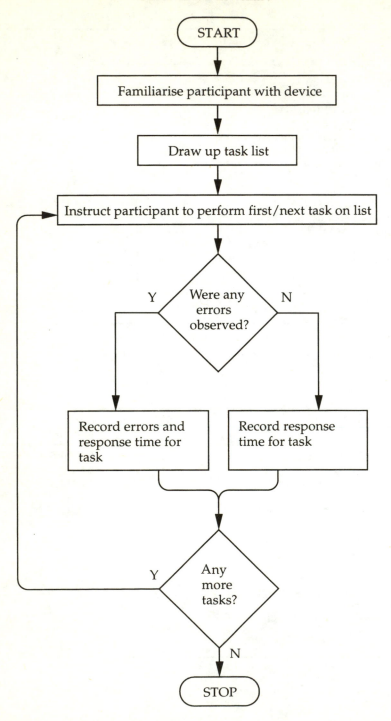

6. QUESTIONNAIRES

Overview

Questionnaires are among the ubiquitous ergonomic methods. They are usually given to a cross-section of the population for purposes of market research, so access to end-users would be advantageous. But in the context of usability evaluations, the questionnaire may be used after a single person has tried the device. They are ideal for accessing quick opinions from target people about usability or other aspects of a product. In that sense, a working form of the product must be in existence.

The questionnaire used in the current project was the system usability scale (SUS), chosen for its brevity and simplistic nature. It is merely a matter of answering ten questions about usability on Likert scales of 1 (strongly disagree with the accompanying statement) to 5 (strongly agree). The questionnaire is then coded and a total score is calculated for usability of that device.

Procedure and advice

The participant (which in this case may be the same person as the analyst) should undertake a thorough user trial with the device in question, executing an exhaustive list of tasks. Having completed this, the participant then fills in the SUS questionnaire based on subjective opinion.

For analysis, each item is given a score from 0 to 4. For all odd-numbered items the score is the scale position minus 1. For all even-numbered items the score is 5 minus the scale position. The sum of all scores is multiplied by 2.5 to obtain the overall usability score, which can range from 0 to 100. This procedure is probably best illustrated in the worked example.

Advantages and disadvantages

- ✓ Efficient means of data collection.
- ✓ Very low on resource usage in execution and analysis.
- ✓ Facilitates comparisons between products.
- ✗ Limited output.
- ✗ A very blunt tool.
- ✗ Can only be usefully applied to an existing product.

Mini bibliography

Brooke, J. (1996) SUS: a 'quick and dirty' usability scale, in Jordan, P. W., Thomas, B., Weerdmeester, B. A., and McClelland, I. L. (eds) *Usability Evaluation in Industry,* London: Taylor & Francis, pp. 189–94.

Jordan, P. W. (1998) *An Introduction to Usability,* London: Taylor & Francis.

Kirakowski, J. (1996) The software usability measurement inventory: background and usage, in Jordan, P. W., Thomas, B., Weerdmeester, B. A. and McClelland, I. L. (eds) *Usability Evaluation in Industry,* London: Taylor & Francis, pp. 169–77.

Kirwan, B. and Ainsworth, L. K. (eds) (1992) *A Guide to Task Analysis,* London: Taylor & Francis.

Oppenheim, A. N. (1992) *Questionnaire Design, Interviewing and Attittude Measurement,* 2nd edn, London: Pinter.

Salvendy, G. and Carayon, P. (1997) Data collection and evaluation of outcome measures, in Salvendy, G. (ed.) *Handbook of Human Factors and Ergonomics,* New York: John Wiley, pp. 1451–70.

Sinclair, M. A. (1990) Subjective assessment, in Wilson, J. R. and Corlett, E. N. (eds) *Evaluation of Human Work: A Practical Ergonomics Methodology*, 2nd edn, London: Taylor & Francis, pp. 69–100.

Questionnaires: pros and cons

Reliability/Validity

Both intra-rater reliability and predictive validity failed to reach significance for questionnaires. Furthermore, on the first use of questionnaires there is a significant difference between the predicted and actual scores. Inter-rater reliability improves to a moderate value on the second application trial.

Resources ✱✱✱

Questionnaires are undoubtedly the quickest method to train and apply. Although this is probably a general result, it may be due in part to the particular questionnaire used in this study, the system usability scale (SUS). SUS is a ten-item questionnaire, so it is inherently quick to use.

Usability ✱✱✱

The brevity of this questionnaire contributes to its high resource usage rating – the best of all the methods. Consistency ratings were better than average on the first trial and improved to a very high level on the second trial.

Efficacy

Questionnaires could generally be applied at any point during the design cycle. SUS, as used in this study, is only really practical with an established product. Access to end-users is preferred.

Examples

Ford radio

| | Strongly disagree | | | | Strongly agree |

1. I think that I would like to use this system frequently

1	2	3	4	5
				✓

2. I found the system unnecessarily complex

1	2	3	4	5
	✓			

3. I thought the system was easy to use

1	2	3	4	5
				✓

4. I think that I would need the support of a technical person to be able to use this system

1	2	3	4	5
✓				

5. I found the various functions in this system were well integrated

1	2	3	4	5
			✓	

6. I thought there was too much inconsistency in this system

1	2	3	4	5
✓				

7. I would imagine that most people would learn to use this system very quickly

1	2	3	4	5
			✓	

8. I found the system very cumbersome to use

1	2	3	4	5
	✓			

9. I felt very confident using the system

1	2	3	4	5
				✓

10. I needed to learn a lot of things before I could get going with this system

1	2	3	4	5
		✓		

Ford radio cont'd

CORING

Odd-numbered items score = scale position − 1	Even-numbered items score = 5 − scale position
Item 1 5 − 1 = 4	Item 2 5 − 2 = 3
Item 3 5 − 1 = 4	Item 4 5 − 1 = 4
Item 5 4 − 1 = 3	Item 6 5 − 1 = 4
Item 7 4 − 1 = 3	Item 8 5 − 2 = 3
Item 9 5 − 1 = 4	Item 10 5 − 3 = 2

Total for odd-numbered items   = 18
Total for even-numbered items   = 16
Grand total   = 34
SUS overall usability score   = grand total × 2.5
                                = 34 × 2.5
                                = 85

Examples

Sharp radio

Strongly disagree   Strongly agree

1. I think that I would like to use this system frequently

1	2	3 ✓	4	5

2. I found the system unnecessarily complex

1 ✓	2	3	4	5

3. I thought the system was easy to use

1	2	3	4	5 ✓

4. I think that I would need the support of a technical person to be able to use this system

1 ✓	2	3	4	5

5. I found the various functions in this system were well integrated

1	2	3 ✓	4	5

Sharp radio cont'd

	Strongly disagree				Strongly agree

6. I thought there was too much inconsistency in this system

1	2 ✓	3	4	5

7. I would imagine that most people would learn to use this system very quickly

1	2	3	4	5 ✓

8. I found the system very cumbersome to use

1 ✓	2	3	4	5

9. I felt very confident using the system

1	2	3	4 ✓	5

10. I needed to learn a lot of things before I could get going with this system

1 ✓	2	3	4	5

SCORING

Odd-numbered items score = scale position − 1	Even-numbered items score = 5 − scale position
Item 1 4 − 1 = 3	Item 2 5 − 1 = 4
Item 3 5 − 1 = 4	Item 4 5 − 1 = 4
Item 5 3 − 1 = 2	Item 6 5 − 2 = 3
Item 7 5 − 1 = 4	Item 8 5 − 1 = 4
Item 9 4 − 1 = 3	Item 10 5 − 1 = 4

Total for odd-numbered items = 16
Total for even-numbered items = 19
Grand total = 35
SUS overall usability score = grand total × 2.5
= 35 × 2.5
= 87.5

Questionnaires flowchart

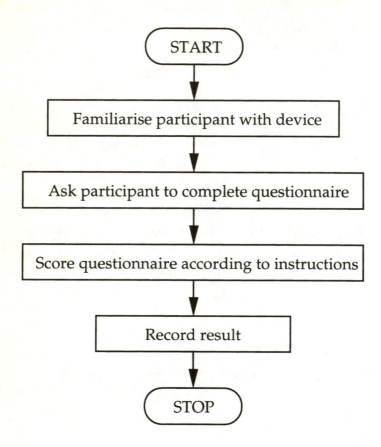

7. HIERARCHICAL TASK ANALYSIS (HTA)

Overview

Hierarchical task analysis, as the name implies, breaks down the task under analysis into a hierarchy of goals, operations and plans. Goals are the unobservable task goals associated with operation of the device. Operations are the observable behaviours or activities which accomplish the goals. Plans are unobservable decisions and planning on behalf of the operator.

The task is described by a task statement, which states the overall goal of the task. This forms the top level of the hierarchy, which is then decomposed into subgoals. Subgoals can be decomposed further until an appropriate stopping point is reached.

The subgoals at any level of the hierarchy must completely describe the superordinate goals; conversely, a superordinate goal must be exhaustively described by its subgoals.

Plans are inserted between levels to provide structure and order to the subtasks immediately below them. Essentially, a plan describes the way in which the subtasks combine to form the superordinate task. Thus plans are very important elements of HTA.

It is useful, although not essential, to have access to the specifications for the design in question.

Procedure and advice

Before even defining the task statement, it is useful to specify the terminology you are going to use. For instance, choose a few words to describe actions, then use these words (as appropriate) to describe all actions in the analysis.

The overall goal is specified at the top of the hierarchy. Break this down into four or five meaningful subgoals, which together make up the overall goal. Break these down further until an appropriate stopping point is reached.

To determine a stopping point, some people use a $p \times c$ criterion. That is, given the current level of description, does the probability p of failure at that task, multiplied by the consequences c of failure, exceed some predetermined criterion? Usually this will not involve a formal calculation, but will be more of an informed judgement on the analyst's part.

The bottom level of any branch will usually be an operation. Everything above has been specifying goals, whereas operations actually say what should be done. They are therefore actions to be made by the operator.

Once all the subgoals have been fully described, the plans should be added. Plans are the 'glue' which dictate how the goals are achieved, and are contingent on current conditions in the environment. For example, a simple plan will say, Do 1 then 2 then 3. And once the plan is finished, the actor returns to the superordinate level. Plans do not have to be linear, and it may be useful to apply boolean logic to them, e.g. Do 1 OR 2 AND 3. Use parentheses to indicate priority.

A complete diagram of goals, subgoals, operations and plans makes up an HTA. This can be tabulated if an alternative representation is required. Tables can be less cumbersome than a diagram if the analysis is particularly large.

51

Advantages and disadvantages

✓ Easily implemented, once the initial concepts have been understood.

✓ Rapid execution, particularly for the first draft; this provides user satisfaction, as good progress is made in little time.

✓ The handbook (Patrick, Spurgeon and Shepherd, 1986) is an invaluable aid to HTA.

✗ Provides more descriptive information than analytical information (though the process of performing an HTA is enlightening in itself).

✗ There is little in the HTA which can be used to directly provide design solutions; such information is necessarily inferred.

✗ HTA does not handle cognitive components of tasks (e.g. decision making), only observable elements.

Mini bibliography

Annett, J., Duncan, K. D., Stammers, R. B. and Gray, M. J. (1971) *Task Analysis*, Department of Employment Training Information Paper 6.

Kirwan, B. (1994) *A Guide to Practical Human Reliability Assessment*, London: Taylor & Francis.

Kirwan, B. and Ainsworth, L. K. (eds) (1992) *A Guide to Task Analysis*, London: Taylor & Francis.

Patrick, J., Spurgeon, P. and Shepherd, A. (1986) *A Guide to Task Analysis: Applications of Hierarchical Methods*, Occupational Services Publications.

Shepherd, A. (1989) Analysis and training in information technology tasks, in Diaper, D. (ed.) *Task Analysis for Human–Computer Interaction,* Chichester: Ellis Horwood, pp. 15–55.

Stammers, R. B. (1996) Hierarchical task analysis: an overview, in Jordan, P. W., Thomas, B., Weerdmeester, B. A. and McClelland, I. L. (eds) *Usability Evaluation in Industry*, London: Taylor & Francis, pp. 207–13.

Stammers, R. B. and Shepherd, A. (1990) Task analysis, in Wilson, J. R. and Corlett, E. N. (eds) *Evaluation of Human Work: A Practical Ergonomics Methodology,* 2nd edn, London: Taylor & Francis, pp. 144–68.

Hierarchical Task Analysis (HTA): pros and cons

Reliability/Validity ✷

Reliability figures for HTA were less than pleasing, although predictive validity was quite strong. However, as reliability and validity are interrelated concepts, it is difficult to accept the validity of this technique without the associated reliability to support it.

Resources ✷

HTA is the most time-intensive method in training and practice. Moreover, execution time is one of the highest, although this considerably improves on the second trial.

Usability ✷✷

One of the better techniques in terms of rated consistency. Resource usage was intriguing, being rated as the second most intensive at trial 1; on trial 2 this had improved to bring it in line with the other methods.

Efficacy

An HTA could be constructed for a concept design, deriving the basics from previous generations. To complete the specifics of the HTA, however, a more concrete design would need to be specified (although this does not have to be in hard-built form).

Examples

Ford radio: diagram format

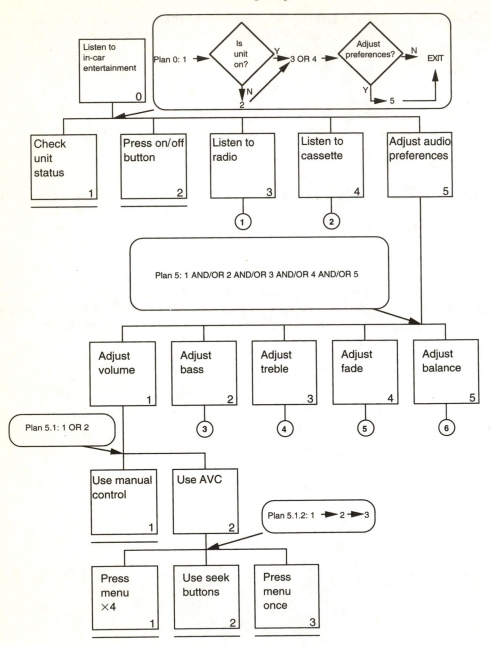

Ford radio: diagram format cont'd

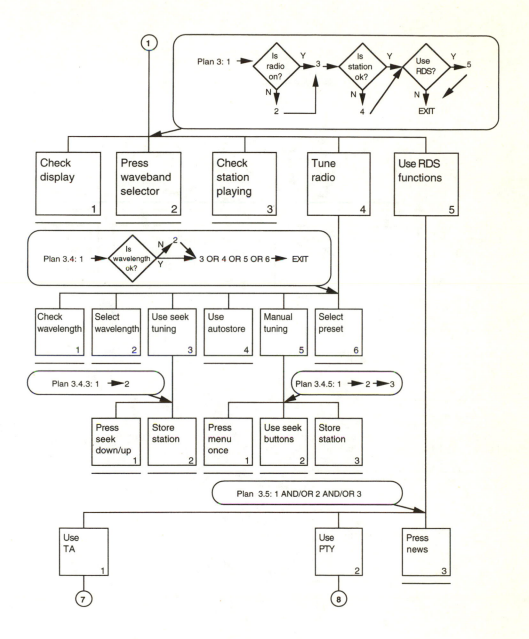

Ford radio: diagram format cont'd

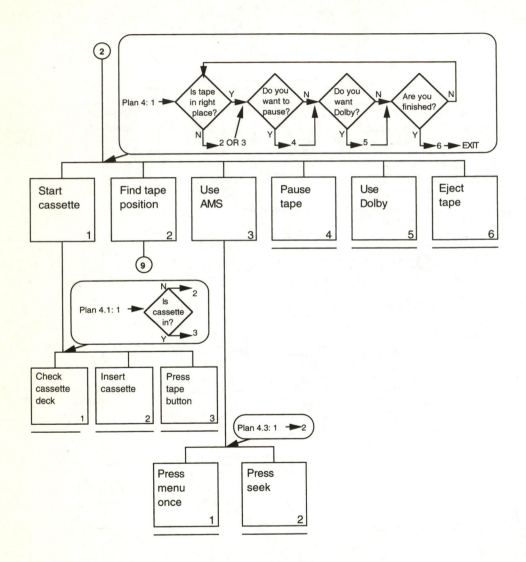

Ford radio: diagram format cont'd

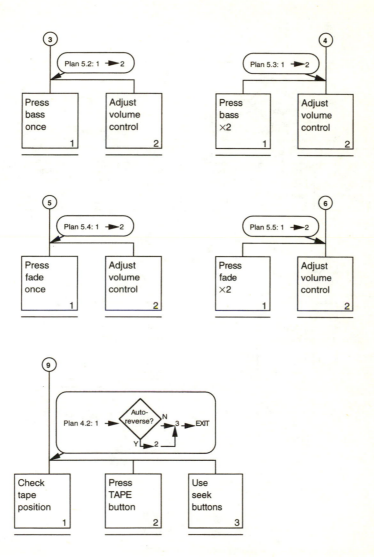

Ford radio: diagram format cont'd

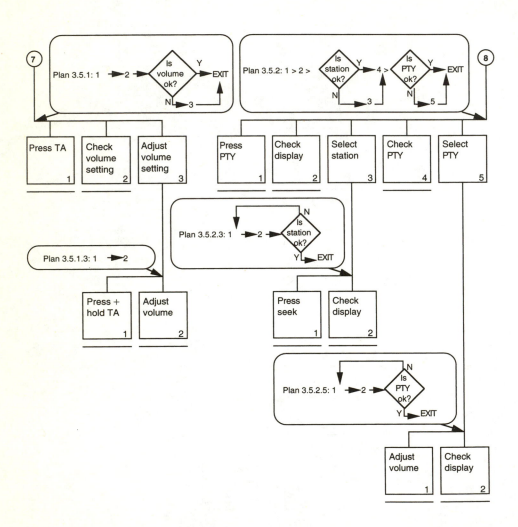

Ford radio: tabular format (abridged)

Super ordinate	Subtask/plan	Notes
0	**LISTEN TO IN-CAR ENTERTAINMENT** P0 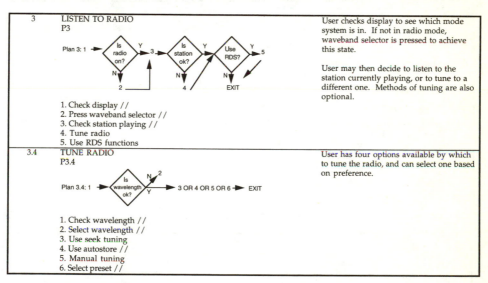 1. Check unit status // 2. Press on/off button // 3. Listen to radio 4. Listen to cassette 5. Adjust audio preferences	User checks whether radio is already on. If not, user switches on by *pressing* on/off knob. User may then decide to listen to radio or cassette, and may wish to adjust their audio preferences.
3	**LISTEN TO RADIO** P3 1. Check display // 2. Press waveband selector // 3. Check station playing // 4. Tune radio 5. Use RDS functions	User checks display to see which mode system is in. If not in radio mode, waveband selector is pressed to achieve this state. User may then decide to listen to the station currently playing, or to tune to a different one. Methods of tuning are also optional.
3.4	**TUNE RADIO** P3.4 1. Check wavelength // 2. Select wavelength // 3. Use seek tuning 4. Use autostore // 5. Manual tuning 6. Select preset //	User has four options available by which to tune the radio, and can select one based on preference.
3.4.3	**USE SEEK TUNING** P3.4.3 Plan 3.4.3: 1 ➤ 2 1. Press seek down/up // 2. Store station //	
3.4.5	**MANUAL TUNING** P3.4.5 Plan 3.4.5: 1 ➤ 2 ➤ 3 1. Press menu once // 2. Use seek buttons // 3. Store station //	
3.5	**USE RDS FUNCTIONS** P3.5 Plan 3.5: 1 AND/OR 2 AND/OR 3 1. Use TA 2. Use PTY 3. Press news //	User has any or all of these RDS options available.

Ford radio: tabular format cont'd

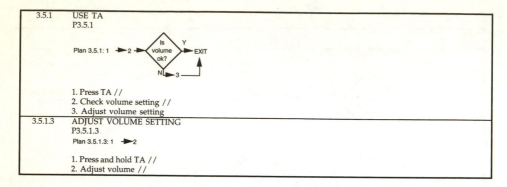

3.5.1	USE TA
	P3.5.1

Plan 3.5.1: 1 → 2 → Is volume ok? — Y → EXIT; N → 3 →

1. Press TA //
2. Check volume setting //
3. Adjust volume setting

3.5.1.3	ADJUST VOLUME SETTING
	P3.5.1.3

Plan 3.5.1.3: 1 → 2

1. Press and hold TA //
2. Adjust volume //

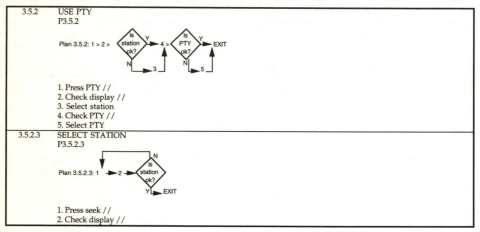

3.5.2	USE PTY
	P3.5.2

Plan 3.5.2: 1 > 2 > Is station ok? — Y → 4 >; N → 3 → ; Is PTY ok? — Y → EXIT; N → 5 →

1. Press PTY //
2. Check display //
3. Select station
4. Check PTY //
5. Select PTY

3.5.2.3	SELECT STATION
	P3.5.2.3

Plan 3.5.2.3: 1 → 2 → Is station ok? — N; Y → EXIT

1. Press seek //
2. Check display //

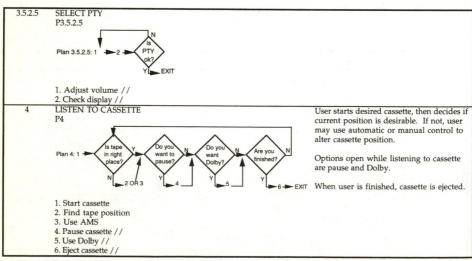

3.5.2.5	SELECT PTY
	P3.5.2.5

Plan 3.5.2.5: 1 → 2 → Is PTY ok? — N; Y → EXIT

1. Adjust volume //
2. Check display //

4	LISTEN TO CASSETTE
	P4

Plan 4: 1 → Is tape in right place? — Y; N → 2 OR 3; Do you want to pause? — N; Y → 4; Do you want Dolby? — N; Y → 5; Are you finished? — N; Y → 6 → EXIT

1. Start cassette
2. Find tape position
3. Use AMS
4. Pause cassette //
5. Use Dolby //
6. Eject cassette //

User starts desired cassette, then decides if current position is desirable. If not, user may use automatic or manual control to alter cassette position.

Options open while listening to cassette are pause and Dolby.

When user is finished, cassette is ejected.

Ford radio: tabular format cont'd

4.1	START CASSETTE P4.1	
	Plan 4.1: 1 → Is cassette in? N→2 Y→3	
	1. Check cassette deck // 2. Insert cassette // 3. Press tape button //	
4.2	FIND TAPE POSITION P4.2	
	Plan 4.2: 1 → Auto-reverse? N→3→EXIT Y→2	
	1. Check tape position // 2. Press tape button // 3. Use seek buttons //	
4.3	USE AMS P4.3	
	Plan 4.3: 1 → 2	
	1. Press menu once // 2. Press seek //	

5	ADJUST AUDIO PREFERENCES P5	User may decide to adjust any or all of these preferences.
	Plan 5: 1 AND/OR 2 AND/OR 3 AND/OR 4 AND/OR 5	
	1. Adjust volume 2. Adjust bass 3. Adjust treble 4. Adjust fade 5. Adjust balance	
5.1	ADJUST VOLUME P5.1	User can adjust volume manually or exploit the automatic volume control on this unit.
	Plan 5.1: 1 OR 2	
	1. Use manual control // 2. Use AVC	
5.1.2	USE AVC P5.1.2	
	Plan 5.1.2: 1 → 2 → 3	
	1. Press menu four times // 2. Use seek buttons // 3. Press menu once //	
5.2	ADJUST BASS P5.2	
	Plan 5.2: 1 → 2	
	1. Press bass once // 2. Adjust volume control //	

5.3	ADJUST TREBLE P5.3	
	Plan 5.3: 1 → 2	
	1. Press bass twice // 2. Adjust volume control //	
5.4	ADJUST FADE P5.4	
	Plan 5.4: 1 → 2	
	1. Press fade once // 2. Adjust volume control //	
5.5	ADJUST BALANCE P5.5	
	Plan 5.5: 1 → 2	
	1. Press fade twice // 2. Adjust volume control //	

Sharp radio: diagram format

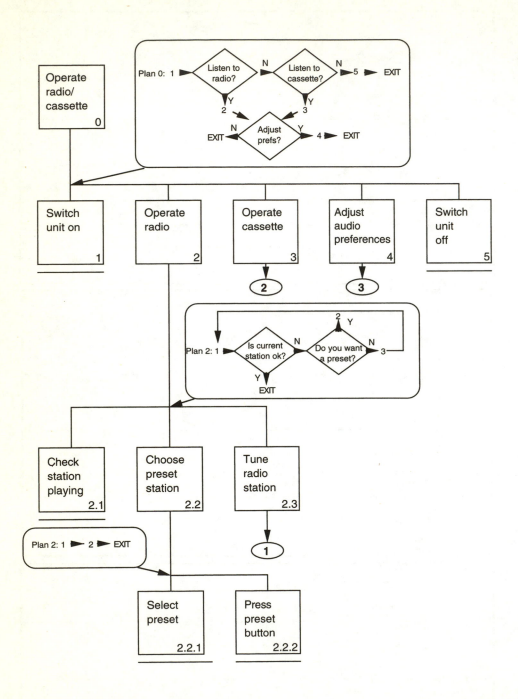

Sharp radio: diagram format cont'd

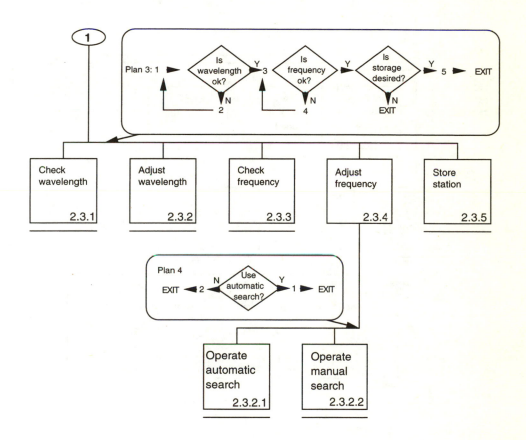

Sharp radio: diagram format cont'd

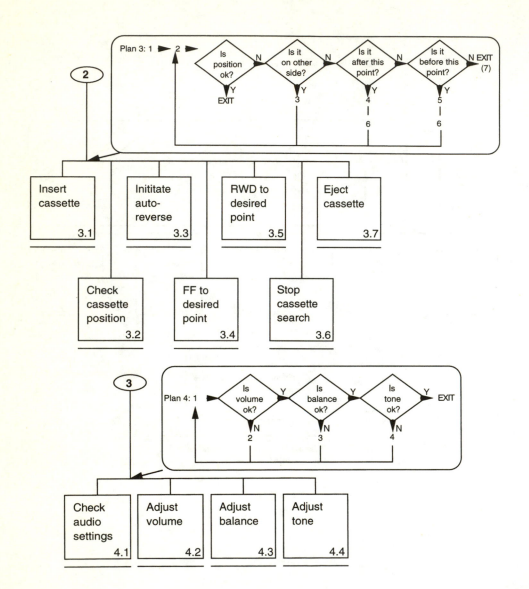

Sharp radio: tabular format

Super ordinate	Subtask/plan	Notes
0	**OPERATE RADIO/CASSETTE** P0 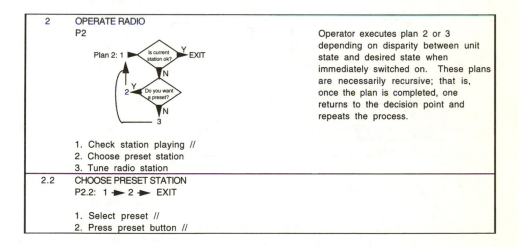 1. Switch unit on // 2. Operate radio 3. Operate cassette 4. Adjust audio preferences 5. Switch unit off //	Operator executes plan 2 or 3 depending on whether radio or cassette listening is desired. Plan 4 is executed only if current audio settings are not within desired parameters.
2	**OPERATE RADIO** P2 1. Check station playing // 2. Choose preset station 3. Tune radio station	Operator executes plan 2 or 3 depending on disparity between unit state and desired state when immediately switched on. These plans are necessarily recursive; that is, once the plan is completed, one returns to the decision point and repeats the process.
2.2	**CHOOSE PRESET STATION** P2.2: 1 ▶ 2 ▶ EXIT 1. Select preset // 2. Press preset button //	

Sharp radio: tabular format cont'd

2.3	**TUNE RADIO STATION** P2.3 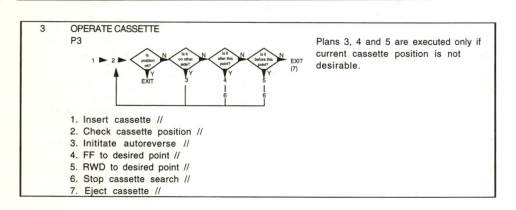 1. Check wavelength // 2. Adjust wavelength // 3. Check frequency // 4. Adjust frequency 5. Store station //	Plans 2, 4 and 5 are optional depending on whether these functions are desired; these plans are also recursive.
2.3.3	**ADJUST FREQUENCY** P2.3.3 1. Operate automatic search // 2. Operate manual search //	Operator decides whether to use automatic search facility or manual search facility.

3	**OPERATE CASSETTE** P3 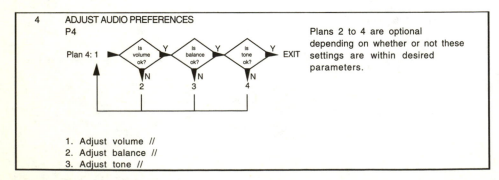 1. Insert cassette // 2. Check cassette position // 3. Inititate autoreverse // 4. FF to desired point // 5. RWD to desired point // 6. Stop cassette search // 7. Eject cassette //	Plans 3, 4 and 5 are executed only if current cassette position is not desirable.

4	**ADJUST AUDIO PREFERENCES** P4 1. Adjust volume // 2. Adjust balance // 3. Adjust tone //	Plans 2 to 4 are optional depending on whether or not these settings are within desired parameters.

HTA flowchart

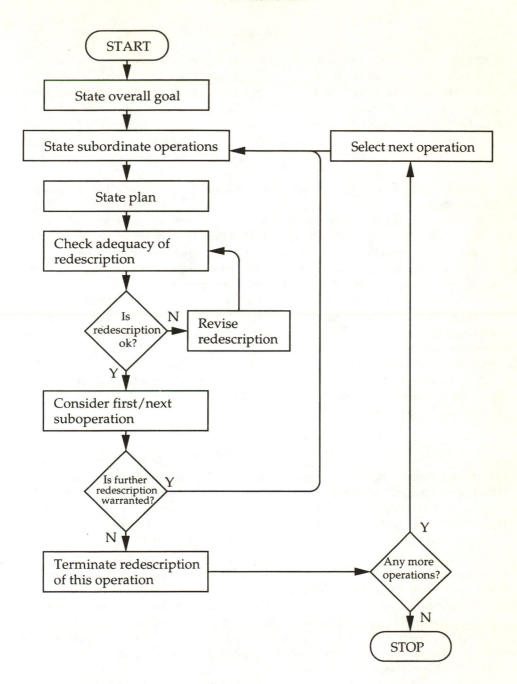

Source: Adapted from Shepherd (1989).

8. REPERTORY GRIDS

Overview

The repertory grid was originally devised as a means of analysing personality. The approach is intended to access an individual's personal view of the world. In that spirit, repertory grids have more recently been applied to product evaluation (Baber, 1996). Thus in this domain the repertory grid may access people's perception of products and may also reveal information about consumer behaviour. The process is different from an Interview in that the intention is to access users' subconscious thoughts about the product. It can therefore be a useful market research tool, used before concept design to determine what consumers find important in a product, or used with an existing product to find out what people think of it. Access to end-users is thus desirable but not essential.

Ideally there should be two people involved, a potential consumer and an interviewer to elicit the consumer's perceptions. However, it is not unreasonable in a product usability assessment for one person to act as both analyst and consumer.

Analysis proceeds by first defining a number of *elements*. In this case, elements may be different types of product, or more likely different forms of the same product. Following this, the method of triads is used to elicit a number of *constructs* about these elements. Complementing constructs are their logical opposites. Thus a grid is formed of elements, constructs and opposites. The grid is completed by filling in ratings for each element on each construct. This quantification is usually analysed by factor analysis; however, Baber (1996) has proposed an alternative method.

It is argued that the repertory grid can provide an insight into product evaluation from the consumer's viewpoint, in addition to supplying a vocabulary for describing and evaluating products.

Procedure and advice

The initial stage of the repertory grid analysis will depend on the form of the product. If it is a prospective analysis, trying to find out what consumers want in a new product, then a briefing about the intended product will be appropriate. On the other hand, if an existing product is being assessed, a thorough user trial will be necessary.

The first stage proper is the definition of elements about the product. Elements are examples of the product in question, so the participant should think up about half a dozen examples with which they are familiar. It is also useful to include hypothetical 'best' and 'worst' examples in this list.

Next constructs are formed by taking sets of three elements at a time (triads). The idea is to identify commonalities between two of the elements that exclude the third element. Consider a car radio. Elements A and B have push-button on/off controls, whereas element C has a twist knob. So 'push-button on/off' is a construct on this triad. The logical opposites of each construct must also be defined; in this example it could be 'twist on/off'. Continue forming constructs on this triad until all are exhausted, then move on to the next triad.

When a list of elements and a list of constructs is complete, the repertory grid can be filled in. With elements as column headers, and constructs as row headers, numbers are entered into the cells representing how much an element is like the construct or its opposite. For the construct 'push-button on/off', elements A and B would rate a 5 (very much like this construct), and element C would rate a 1 (very much like the opposite).

Analysing a repertory grid may be either qualitative or quantitative. By examining responses to the participant's hypothetical best product, it is possible to determine which qualities are important for user acceptance, and to see how the current design measures up to them. If more concrete numbers are required, factor analysis can be used to quantify these results (an explanation of factor analysis can be found in most decent statistical texts).

Advantages and disadvantages

✓ Potentially useful product evaluation technique, providing insight into actual consumer perceptions.
✓ Not a difficult technique to execute.
✓ Can provide useful information for designers.
✗ Cluster analysis is complex and time-consuming.
✗ Necessity of an independent participant for grid construction.
✗ Not an explicitly predictive technique.

Mini bibliography

Baber, C. (1996) Repertory grid theory and its application to product evaluation, in Jordan, P. W., Thomas, B., Weerdmeester, B. A. and McClelland, I. L. (eds) *Usability in Industry*, London: Taylor & Francis, pp. 157–65.

Fransella, F. and Bannister, D. (1977) *A Manual for Repertory Grid Technique*, London: Academic Press.

Kelly, G. A. (1955) *The Psychology of Personal Constructs*, New York: Norton.

Oosterwegel, A. and Wickland, R. A. (1995) *The Self in European and North American Culture: Development and Processes*, Dordrecht: Kluwer Academic.

Repertory Grids: pros and cons

Reliability/Validity ✱

Although predictive validity for repertory grids was not too bad, both measures of reliability were disappointing. Therefore the validity measure must be treated with caution.

Resources ✱✱

Repertory grids took a moderate amount of time to train participants in, with practice times being higher (though not nearly so great as the previous three methods). Execution time was again moderate, but this improved considerably on the second application.

Usability ✶✶✶

Ratings of consistency were average on the first trial, improving on the second trial to put it on a par with the other methods. Resource usage ratings place this method among the best. Remember this technique requires two participants.

Efficacy

Output from a repertory grid provides designers with insight into what consumers find important in a product. Therefore this method will be best exploited very early on in the design cycle, most appropriately at the concept stage.

Example

This example presents responses to a repertory grid analysis involving both types of car radio as elements, plus a hypothetical worst radio and a hypothetical best radio, to gauge whether a construct is perceived as good or bad.

Constructs	Elements				Opposites
	Sharp	Ford	Worst	Best	
Mode dependent	1	5	5	1	Separate functions
Push-button operation	2	5	1	5	Knob-turn operation
Bad labelling	2	5	5	1	Clear labelling
Easy controls	1	5	1	5	Fiddly controls
Poor functional grouping	4	5	5	1	Good functional grouping
Good illumination	2	4	1	5	Poor illumination

5 = left side very much applicable (right side not applicable at all)
4 = left side somewhat applicable (right side not really applicable)
3 = in between
2 = left side not really applicable (right side somewhat applicable)
1 = left side not applicable at all (right side very much applicable)
0 = characteristic irrelevant

Repertory grids flowchart

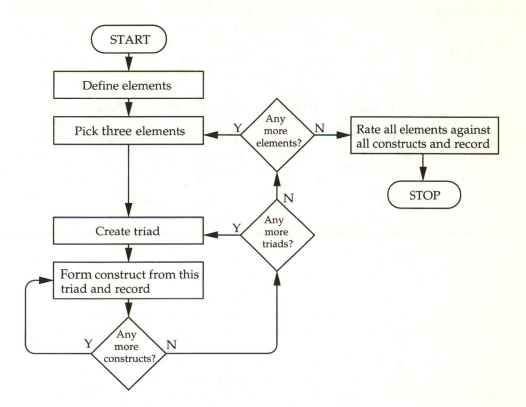

9. TASK ANALYSIS FOR ERROR IDENTIFICATION (TAFEI)

Overview

TAFEI attempts to predict errors with device use by modelling the interaction between user and device. It assumes that people use devices in a purposeful manner, such that the interaction may be described as a 'cooperative endeavour', and it is by this process that problems arise. Furthermore, the technique makes the assumption that actions are constrained by the state of the product at any particular point in the interaction, and that the device offers information to the user about functionality.

Procedurally, TAFEI is comprised of three main stages:

1. A hierarchical task analysis (HTA) is performed to model the human side of the interaction.
2. State space diagrams (SSDs) are constructed to represent the behaviour of the artifact; plans from the HTA are mapped onto the SSD to form the TAFEI diagram.
3. A transition matrix is devised to display state transitions during device use.

TAFEI aims to assist the design of artifacts by illustrating when a state transition is possible but undesirable (i.e. illegal). Making all illegal transitions impossible should facilitate the cooperative endeavour of device use.

Procedure and advice

The first step in a TAFEI analysis is to obtain an appropriate HTA for the device. As TAFEI is best applied to scenario analyses, it might be wise to consider just one specific portion of the HTA (e.g. a specific, closed-loop task of interest) rather than the whole design. Once this is done, the analysis proceeds to constructing state space diagrams (SSDs) for device operation.

An SSD essentially consists of lists of states which the device can be in. For each list, there will be a current state (at the top of the list) and a list of possible exit conditions to take it to other states. At a very basic level, the current state might be 'off', with the exit condition 'switch on' taking the device to the state 'on'. On completing the SSD, it is very important to have an exhaustive set of states for the device under analysis.

Numbered plans from the HTA are then mapped onto the SSD, indicating which human actions take the device from one state to another. Thus the plans are mapped onto the state transitions (if a transition is activated by the machine, this is also indicated on the SSD). This produces a TAFEI diagram.

From the viewpoint of improving usability, the most important part of the analysis is the transition matrix. All possible states are entered as headers on a matrix. The cells represent state transitions (e.g. the cell at row 1, column 2 represents the transition between state 1 and state 2) and are then filled in one of three ways:

- If a transition is deemed impossible (i.e. you simply cannot go from this state to that one), enter a dash (–) into the cell.
- If a transition is deemed possible and desirable (i.e. it progresses the user towards the goal state – a correct action) this is a legal transition, so enter L into the cell.
- If a transition is both possible but undesirable (a deviation from the intended path – an error) this is an illegal transition, so enter I into the cell.

The idea behind TAFEI is that usability may be improved by making all illegal transitions (errors) impossible, thereby limiting the user to only performing desirable actions. It is up to the analyst to conceive design solutions that achieve this.

Advantages and disadvantages

✓ Structured and thorough procedure.
✓ Sound theoretical underpinning.
✓ Flexible, generic methodology.
✗ Not a rapid technique, as HTA and SSD are prerequisites.
✗ Validation research currently sparse.
✗ Limited to linear tasks.

Mini bibliography

Baber, C. and Stanton, N. A. (1994) 'Task analysis for error identification: a methodology for designing error-tolerant consumer products', *Ergonomics*, **37**, pp. 1923–41.

Stanton, N. A. and Baber, C. (1996a) A systems approach to human error identification. *Safety Science*, **22**, pp. 215-28.

Stanton, N. A. and Baber, C. (1996b) Task analysis for error identification: applying HEI to product design and evaluation, in Jordan, P. W., Thomas, B., Weerdmeester, B. A. and McClelland, I. L. (eds) *Usability Evaluation in Industry,* London: Taylor & Francis, (pp. 215–24).

Stanton, N. A. and Baber, C. (1998) A systems analysis of consumer products, in Stanton, N. A. (ed.) *Human Factors in Consumer Products*, London: Taylor & Francis, pp. 75–90.

Task Analysis For Error Identification (TAFEI): pros and cons

Reliability/Validity

Only a small proportion of our analysts completed this technique, so it was not possible to compute reliability statistics. Although the validity measures were respectable, the small sample size means they are tentative.

Resources *

TAFEI was second only to HTA in training and practice time. Rehearsal time was more than twice that taken for training. This method required by far the greatest execution time on both occasions; the majority of participants failed to complete the analysis within the time allotted.

Usability **

Our analysts rated TAFEI as moderately consistent; it was by no means the worst on either trial. Ratings of resource usage were clearly worse than all other methods, although perceptions had improved considerably by trial 2.

Efficacy

With HTA being a prerequisite, TAFEI would only work with a formalised design in existence. As with PHEA, the error output would be of most use around the prototyping stage.

Examples

Ford radio: TAFEI diagram

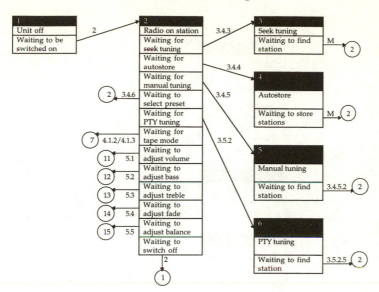

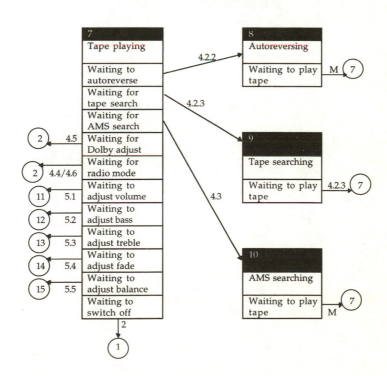

Ford radio: TAFE I diagram cont'd

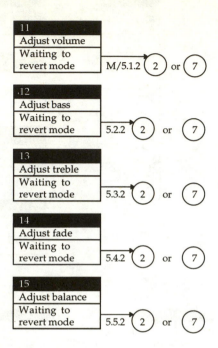

Ford radio: transition matrix

	1	2	3	4	5	6	7	8	9	10	11	12	13	14	15
1	–	L	–	–	–	–	–	–	–	–	–	–	–	–	–
2	L	–	L	L	L	L	L	–	–	–	L	L	L	L	L
3	I	L	–	I	I	I	I	–	–	–	I	I	I	I	I
4	I	L	I	–	I	I	I	–	–	–	I	I	I	I	I
5	I	L	–	I	–	I	I	–	–	–	I	I	I	I	I
6	I	L	–	I	–	–	I	–	–	–	I	I	I	I	I
7	L	L	–	–	–	–	–	L	L	L	L	L	L	L	L
8	I	I	–	–	–	–	L	–	–	–	I	I	I	I	I
9	I	–	–	–	–	–	L	I	–	I	I	I	I	I	I
10	I	I	–	–	–	–	L	I	–	–	I	I	I	I	I
11	I	L	–	–	–	–	–	–	–	–	–	–	–	–	–
12	I	L	–	–	–	–	I	–	–	–	–	–	I	I	I
13	I	L	–	–	–	–	I	–	–	–	I	I	–	I	I
14	I	L	–	–	–	–	I	–	–	–	I	I	I	–	I
15	I	L	–	–	–	–	I	–	–	–	I	I	I	I	–

Sharp radio: TAFEI diagram

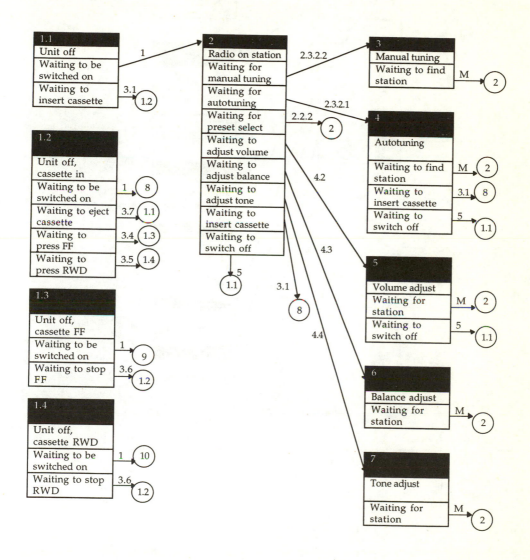

Sharp radio: TAFEI diagram cont'd

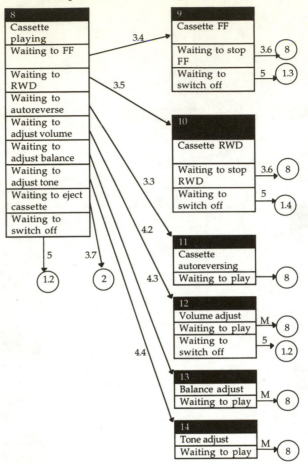

Sharp radio: transition matrix

	1.1	1.2	1.3	1.4	2	3	4	5	6	7	8	9	10	11	12	13	14
1.1	–	I	–	–	L	–	–	–	–	–	–	–	–	–	–	–	–
1.2	L	–	I	I	–	–	–	–	–	–	L	–	–	–	–	–	–
1.3	–	I	–	–	–	–	–	–	–	–	–	L	–	–	–	–	–
1.4	–	I	–	–	–	–	–	–	–	–	–	–	L	–	–	–	–
2	L	–	–	–	–	L	L	L	L	L	L	–	–	–	–	–	–
3	–	–	–	–	L	–	–	–	–	–	–	–	–	–	–	–	–
4	L	–	–	–	L	–	–	–	–	–	L	–	–	–	–	–	–
5	I	–	–	–	L	–	–	–	–	–	–	–	–	–	–	–	–
6	–	–	–	–	L	–	–	–	–	–	–	–	–	–	–	–	–
7	–	–	–	–	L	–	–	–	–	–	–	–	–	–	–	–	–
8	–	I	–	–	L	–	–	–	–	–	–	L	L	L	L	L	L
9	–	–	I	–	–	–	–	–	–	–	L	–	–	–	–	–	–
10	–	–	–	I	–	–	–	–	–	–	L	–	–	–	–	–	–
11	–	–	–	–	–	–	–	–	–	–	L	–	–	–	–	–	–
12	–	I	–	–	–	–	–	–	–	–	L	–	–	–	–	–	–
13	–	–	–	–	–	–	–	–	–	–	L	–	–	–	–	–	–
14	–	–	–	–	–	–	–	–	–	–	L	–	–	–	–	–	–

TAFEI flowchart

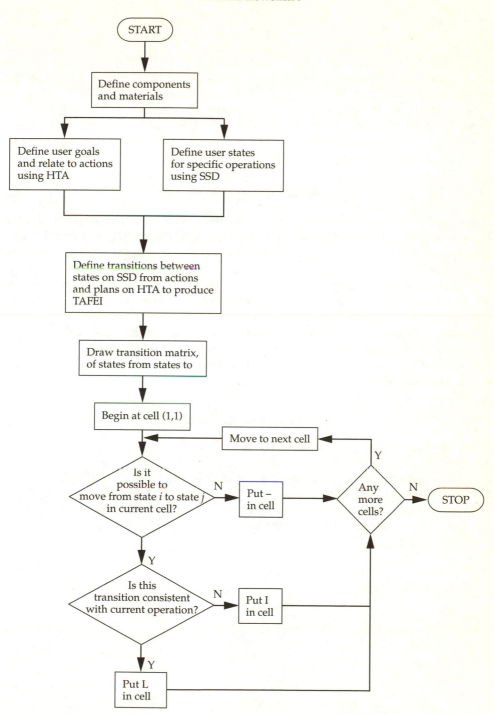

10. LAYOUT ANALYSIS

Overview

Like link analysis, layout analysis is based on spatial diagrams of the product, thus its output directly addresses interface design. It may be used as soon as a concrete design has been conceived.

The method simply analyses an existing design and suggests improvements to the interface arrangement based on functional grouping. The theory is that the interface should mirror the user's structure of the task, and the conception of the interface as a task map greatly facilitates design (Easterby, 1984).

Procedure and advice

Layout analysis starts off by arranging the components of the interface into functional groupings, based on the opinions of the analyst. These groups are then organised by importance of use, sequence of use and frequency of use. That is, the analyst might wish to make the most important group of components most readily available, although this might be tempered by sequence or frequency of use. In a similar manner, the components within each functional group are then reorganised, again according to importance, sequence and frequency of use.

Components within a functional group stay in that group throughout the analysis; they do not move somewhere else in the reorganisation stage.

At the end of this process, the analyst has redesigned the device in accordance with the user's model of the task.

Advantages and disadvantages

✓ Easy to implement in the applied setting.
✓ Low demand on resources.
✓ Tangible output means little training required.
✗ Poor reliability and validity.
✗ Output limited to issues of layout, rather than errors or task times.
✗ Paucity of literature for layout analysis.

Mini bibliography

Easterby, R. (1984) Tasks, processes and display design, in Easterby, R. and Zwaga, H. (eds) *Information Design*, Chichester: John Wiley, pp. 19–36.

Stanton, N. A. and Young, M. S. (1998b) Ergonomics methods in consumer product design and evaluation, in Stanton, N. A. (ed) *Human Factors in Consumer Products*, London: Taylor & Francis, pp. 21–53.

Layout Analysis: pros and cons

Reliability/Validity

Statistics for intra-rater reliability and predictive validity for layout analysis were not impressive. Inter-rater reliability improved to a moderate result on trial 2.

Resources ✳✳✳

Layout analysis is among the quickest techniques to both train and apply. A prepared HTA for the device in question is useful but not essential.

Usability ✳✳✳

Layout analysis scored better than average on measures of consistency and resource usage.

Efficacy

As with link analysis, a layout analysis can be performed at any stage when there is a formalised design. It is probably of most use before the prototyping stage, as its output is geared towards rearranging the interface layout.

Examples

Ford radio

INITIAL DESIGN

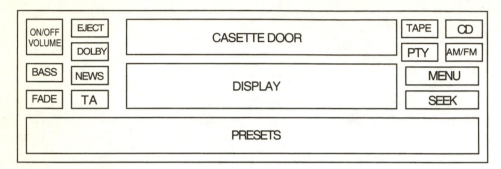

FUNCTIONAL GROUPINGS

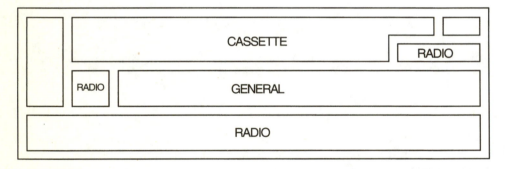

IMPORTANCE OF USE

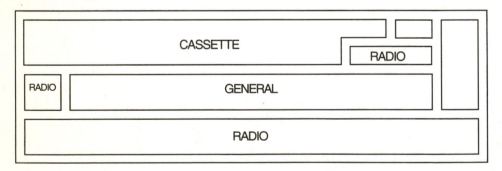

Ford radio cont'd

SEQUENCE OF USE

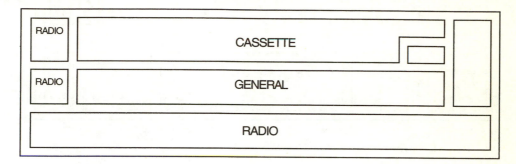

WITHIN FUNCTIONAL GROUPINGS

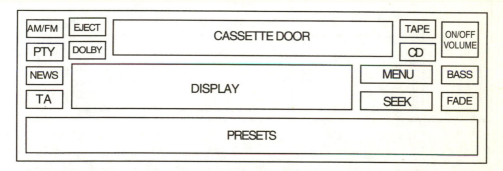

REVISED DESIGN BY IMPORTANCE, FREQUENCY AND SEQUENCE OF USE

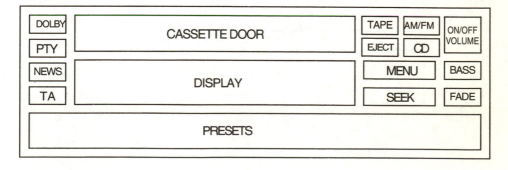

Sharp radio

INITIAL DESIGN

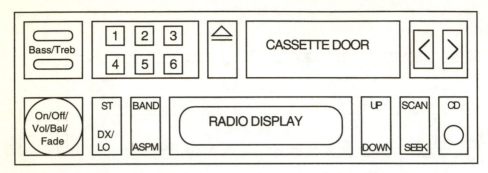

FUNCTIONAL GROUPINGS

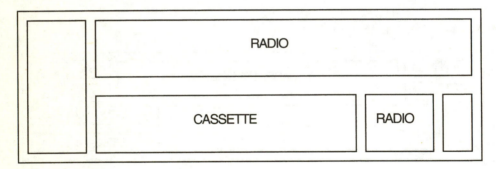

IMPORTANCE OF USE

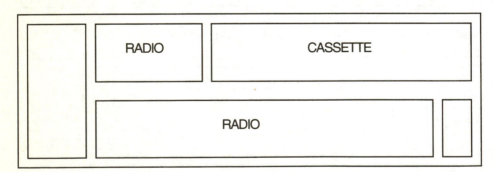

Sharp radio cont'd

SEQUENCE OF USE (UNCHANGED)

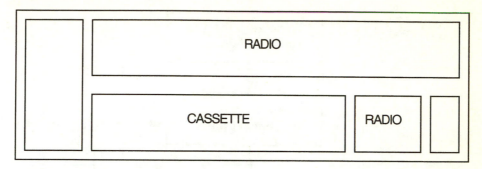

WITHIN FUNCTIONAL GROUPINGS

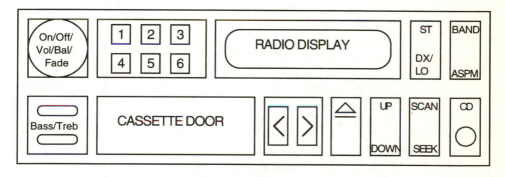

REVISED DESIGN BY IMPORTANCE, FREQUENCY AND SEQUENCE OF USE

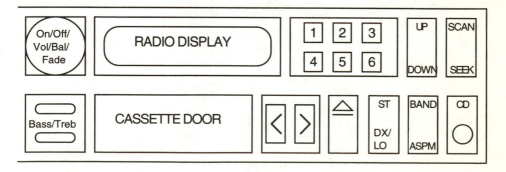

Layout analysis flowchart

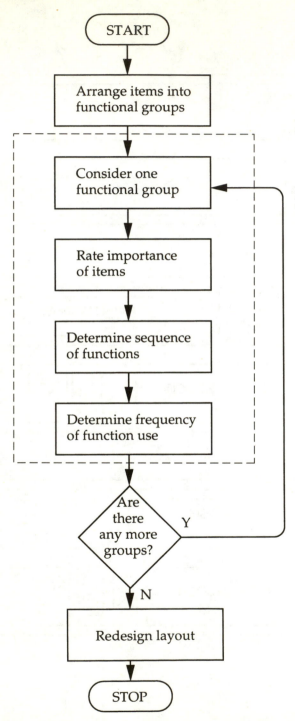

START

Arrange items into functional groups

Consider one functional group

Rate importance of items

Determine sequence of functions

Determine frequency of function use

Are there any more groups?

Y

N

Redesign layout

STOP

11. INTERVIEWS

Overview

.Interviews are general information-gathering exercises, in this context intended to elicit users' or designers' views about a particular task or system. As a result, they possess great flexibility in application, although in usability evaluations a user trial is implied before carrying out an interview. Furthermore, the interview technique is very well documented, with an abundance of literature on this method (Sinclair, 1990). Interviews are generally dyadic, and a working example of the device to be studied would be an advantage (although the interview can be used in market research, the application described here is in assessing usability of a device). The interview may take one of three forms: structured (essentially an orally administered questionnaire); semi-structured (a more flexible approach, with questioning being guided but not restricted by a crib sheet); or unstructured (a free-form discussion).

The main advantage of an interview is its familiarity to the respondent as a technique and this, combined with the face-to-face nature, is likely to elicit more information, and probably more accurate information. In addition, because it is administered, the interviewer can pursue intriguing lines of inquiry. If access to end-users is available, the output might be more revealing by using these people as interviewees.

Procedure and advice

For the analysis of in-car devices, particularly with untrained interviewers, it is probably wise to adopt a semistructured interview format. We suggest using the Ravden and Johnson (1989) checklist as a basis, for this is effectively a ready-made structure. Although this was designed for assessing usability of human–computer interfaces, it seems amenable to interaction with other devices. A semistructured interview is essentially an administered questionnaire, with the advantage of flexibility, both in responses and in having a domain expert (i.e. the interviewer) present.

The interviewee should be granted an exhaustive user trial with the device under analysis, then interviewed for their thoughts. Each section title of the checklist should be used as a prompt for asking questions (e.g. Let's talk about visual clarity – did you think information was clear and well organised?). Note that the structure is just the bones for building an interview around; it is by no means fixed and should not be viewed as a script for asking questions. It is more of an agenda to ensure all aspects are covered. The interviewer should direct the questioning from open questions (What did you think of this aspect?) through probing questions (Why do you think that?) to more closed ones (Is this a good thing?). It may be useful to keep a protocol sheet to hand as a prompt for this. The idea is that the interviewer opens a line of inquiry with an open question, then follows it up. When one line of inquiry is exhausted, the interviewer moves to another line of inquiry. By doing this for every aspect of the device, one can be sure of having conducted a thorough interview. It is helpful to have prepared a data sheet for filling in responses during the interview.

As with checklists, interviews are adaptive, and if the interviewer feels that any particular section is irrelevant, they are free to exclude it. The professional wisdom of the interviewer can be an advantage for this technique.

Advantages and diadvantages

✓ Familiar technique to most respondents.
✓ Flexibility – information can be followed up 'on-line'.
✓ Structured interview offers consistency and thoroughness.
✗ Necessitates a user trial.
✗ Time-consuming analysis.
✗ Demand characteristics of situation may lead to misleading results.

Mini bibliography

Jordan, P. W. (1998) *An Introduction to Usability*, London: Taylor & Francis.
Kirwan, B. & Ainsworth, L. K. (eds) (1992) *A Guide to Task Analysis*, London: Taylor & Francis.
Ravden, S. J. and Johnson, G. I. (1989) *Evaluating Usability of Human–Computer Interfaces: A Practical Method*, Chichester: Ellis Horwood.
Sinclair, M. A. (1990) Subjective assessment, in Wilson, J. R. and Corlett, E. N. (eds) *Evaluation of Human Work: A Practical Ergonomics Methodology*, 2nd edn, London: Taylor & Francis, pp. 69–100.
van Vianen, E., Thomas, B. and van Nieuwkasteele, M. (1996) A combined effort in the standardization of user interface testing, in Jordan, P. W., Thomas, B., Weerdmeester, B. A. and McClelland, I. L. (eds) *Usability Evaluation in Industry*, London: Taylor & Francis, pp. 7–17.
Young, M. S. and Stanton, N. A. (1999) The interview as a usability tool, in Memon, A. and Bull, R. (eds) *Handbook on the Psychology of Interviewing*, Chichester: John Wiley.

Interviews: pros and cons

Reliability/Validity

The structured interview, as it was applied in the present experiment, did not score well on any of the ratings of reliability or validity.

Resources ✱✱

Interviews took slightly longer to train and use than the previous methods, probably due to the fact that the interview technique is a refined process which needs to be understood properly before it can be implemented effectively. The interview implies at least two people are needed to carry out this evaluation.

Usability ✱✱

Average ratings of consistency were assigned to interviews, and only slightly higher than average scores for resource usage. The interview was thus moderately well accepted by our trainees.

Efficacy

The interview can easily be applied at any stage in the design process, from asking people what they want in a device to eliciting opinions about an existing design.

Output is not limited to issues of interface design. Ideally, end-users should be targeted for their views.

Examples

Ford radio

SECTION 1: VISUAL CLARITY
Information displayed on the screen should be clear, well-organised, unambiguous and easy to read

- Display is clear and legible, but could have problems in poor illumination
- Ambiguous abbreviations (e.g. Dolby symbol, PTY)

SECTION 2: CONSISTENCY
The way the system looks and works should be consistent at all times

- Inconsistency in menu modes, when progression is achieved by using seek buttons
- PTY function is odd, using volume knob to choose type, then seek to select

SECTION 3: COMPATIBILITY
The way the system looks and works should be compatible with user conventions and expectations

- RDS and PTY functions required use of the manual before operation
- Autoreverse function incompatible with previous conventions

SECTION 4: INFORMATIVE FEEDBACK
Users should be given clear, informative feedback on where they are in the system, what actions they have taken, whether these actions have been successful and what actions should be taken next

- Feedback is generally very good, obvious what is happening most of the time

SECTION 5: EXPLICITNESS
The way the system works and is structured should be clear to the user

- PTY label is not explicit on initial operation

SECTION 6: APPROPRIATE FUNCTIONALITY
The system should meet the needs and requirements of users when carrying out tasks

- Depressing seek when TA is engaged initiates TP seek – this is annoying
- PTY function is superfluous, especially for the prominence of the control

SECTION 7: FLEXIBILITY AND CONTROL
The interface should be sufficiently flexible in structure, in the way information is presented and in terms of what the user can do, to suit the needs and requirements of all users, and to allow them to feel in control of the system

- The system is extremely flexible and responds more than adequately to user input

SECTION 8: ERROR PREVENTION AND CORRECTION
The system should be designed to minimise the possibility of user error, with in-built facilities for detecting and handling those which do occur; users should be able to check their inputs and to correct errors, or potential error situations, before the input is processed

- The possibility of user error is very small, only minor non-critical errors may occur

SECTION 9: USER GUIDANCE AND SUPPORT
Informative, easy-to-use and relevant guidance and support should be provided, both on the computer (via an online help facility) and in hard-copy document form, to help the user understand and use the system

- Manual is understandable but not well structured
- Specific sections needed for radio, tape and RDS functions, plus a reference section

SECTION 10: SYSTEM USABILITY PROBLEMS

- Menu functions are something of a mystery
- Autoseek only works on FM waveband

SECTION 11: GENERAL SYSTEM USABILITY

- Best aspect: RDS functions are most useful
- Worst aspect: PTY feature is overstated
- Common mistakes: pressing TA initiates TP seek
- Irritating aspects: volume control causes static interference
- Recommended changes: relegate some obscure functions (e.g. PTY) to menu

Sharp radio

SECTION 1: VISUAL CLARITY
Information displayed on the screen should be clear, well-organised, unambiguous and easy to read

- There is a certain amount of visual clutter on the LCD, particularly with respect to preset station number
- Little or no discrimination between functions
- Writing (labelling) is small but readable
- Labelling is all upper case
- Ambiguous abbreviations (e.g. DX/LO, ASPM ME-SCAN)
- Gear lever can obscure vision to controls

SECTION 2: CONSISTENCY
The way the system looks and works should be consistent at all times

- Tuning buttons present inconsistent labelling (especially scan and seek functions)
- Moded functions create problems in knowing how to initiate the function (i.e. press vs. press and hold)

SECTION 3: COMPATIBILITY
The way the system looks and works should be compatible with user conventions and expectations

- Scan and seek buttons lack compatibility
- Four functions on on/off switch makes it somewhat incompatible
- Programming preset stations may not be intuitive for a novice user, but it is compatible with other systems
- Autoreverse function could cause cognitive compatibility problems, particularly when involving FF/RWD functions

SECTION 4: INFORMATIVE FEEDBACK
Users should be given clear, informative feedback on where they are in the system, what actions they have taken, whether these actions have been successful and what actions should be taken next

- Tactile feedback is poor, particularly for the on/off switch
- Instrumental and operational feedback generally good, except in the case of programming a preset station, when operational feedback is poor

SECTION 5: EXPLICITNESS
The way the system works and is structured should be clear to the user

- The novice user may not understand how to program stations without consulting the manual
- Resuming normal cassette playback after FF or RWD is not clear
- Initiating the autoreverse function is not obvious

SECTION 6: APPROPRIATE FUNCTIONALITY
The system should meet the needs and requirements of users when carrying out tasks

- Rotating dial is not appropriate for front/rear fader control; maybe a joystick control would be better
- Prompts for task steps may be useful when programming stations
- Radio is muted when tuning – perhaps it would be possible to monitor the tuning process

SECTION 7: FLEXIBILITY AND CONTROL
The interface should be sufficiently flexible in structure, in the way information is presented and in terms of what the user can do, to suit the needs and requirements of all users, and to allow them to feel in control of the system

- Novice users may experience some difficulty
- Users with larger fingers may find controls fiddly
- Radio is inaudible while winding cassette – this is inflexible

SECTION 8: ERROR PREVENTION AND CORRECTION

The system should be designed to minimise the possibility of user error, with in-built facilities for detecting and handling those which do occur; users should be able to check their inputs and to correct errors, or potential error situations, before the input is processed

- There is no undo function for stored stations
- Separate functions would be better initiated from separate buttons (e.g. tuning up/down)
- Balance/volume control is conducive to errors

SECTION 9: USER GUIDANCE AND SUPPORT

Informative, easy-to-use and relevant guidance and support should be provided, both on the computer (via an online help facility) and in hard-copy document form, to help the user understand and use the system

- Manual is not well structured (no contents page; installation instructions are mixed up with operations)
- Relevant manual sections are not easy to find, but this is alleviated somewhat by the manual being short
- Instructions in the manual are matched to the task

SECTION 10: SYSTEM USABILITY PROBLEMS

- Minor problems in understanding function of two or three buttons
- Finding information in the manual can be problematic
- Writing (labelling) on the radio is small
- Operation of scan button can be misunderstood
- Treble and bass controls are tiny

SECTION 11: GENERAL SYSTEM USABILITY

- Best aspect: this radio is *not* mode-dependent
- Worst aspects: ambiguity in button labelling; tactile feedback on volume control could be improved
- Confusing/difficult aspects: on initial use of FF and RWD when in autoreverse mode
- Irritating aspects: volume control causes static interference
- Common mistakes: adjusting balance instead of volume
- Recommended changes: introduce push-button operation for on/off control

Interviews flowchart

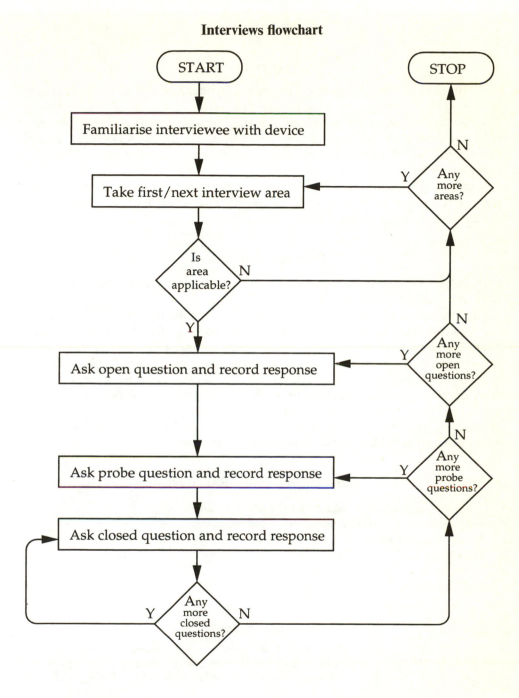

12. HEURISTICS

Overview

A heuristic analysis is quite possibly the simplest technique available. It merely requires the analyst (preferably an expert with the product under review, one with knowledge of its use and misuse) to sit down and use their subjective judgement to decide whether a device is usable, error-inducing, safe and generally well designed. As such, the analysis can be performed on a product in any form (i.e. concept, prototype) and at any stage in the design cycle.

Typical output from a heuristic analysis centres around functionality and labelling, although it may just as easily be used to assess other aspects of usability, and to predict errors and performance times.

Procedure and advice

It would be a contradiction in terms to apply any hard and fast rules to heuristics. However, some advice may be given.

A thorough analysis will result from having spent some time familiarising oneself with the device and any accompanying documentation (e.g. manual). Then the analyst should try to perform an exhaustive set of tasks in order to explore every aspect of device functioning. Notes should be taken throughout on good as well as bad points of usability for feedback into the design process. If it is of interest, the analyst can also assess the documentation for usability issues. Finally, the analyst may wish to offer solutions for any problems encountered.

Care must be exercised during the course of the analysis to be as impartial as possible, even if the analyst has a vested interest in the outcome.

Advantages and disadvantages

✓ Very simple to execute, requiring little previous knowledge.
✓ Very efficient on resources (both time and materials).
✓ Highly usable method.
✗ Highly subjective technique.
✗ Unstructured.
✗ Lacks reliability, comprehensiveness and auditability.

Mini bibliography

de Vries, G., Hartevelt, M. and Oosterholt, R. (1996) Private camera conversation: a new method for eliciting user responses, in Jordan, P. W., Thomas, B., Weerdmeester, B. A. and McClelland, I. L. (eds) *Usability Evaluation in Industry*, London: Taylor & Francis, pp. 147–55.
Kirwan, B. (1992) 'Human error identification in human reliability assessment – part 2', *Applied Ergonomics*, **23**, pp. 371–81.
Nielsen, J. and Molich, R. (1990) Heuristic evaluation of user interfaces, in Chew, J. C. and Whiteside, J. (eds) *Empowering People: CHI '90 Conference Proceedings*, Monterey, CA: ACM Press, pp. 249–56.

Pejtersen, A. M. and Rasmussen, J. (1997) Effectiveness testing of complex systems, in Salvendy, G. (ed.) *Handbook of Human Factors and Ergonomics*, 2nd edn, New York: John Wiley, pp. 1514–42.

Heuristics: pros and cons

Reliability/Validity

The unstructured nature of heuristics leads to poor results when assessing reliability and predictive validity. Statistically speaking, this is not an impressive method.

Resources ✱✱✱

One of the quickest methods to train and use, although first application can be quite time-consuming. There are no prerequisites for use, and heuristics can essentially be performed by one person.

Usability ✱✱

Trained users gave heuristics the lowest consistency rating of all the techniques, although in terms of resource usage it rated as one of the best.

Efficacy

Heuristics, as a general tool, can by definition be applied at any stage of design, and consequently on any form of product. If access to end-users is available, heuristics is probably best applied using a prototype evaluation technique. Output from a heuristic analysis typically covers immediate interface design (i.e. labelling, functionality).

Examples

Ford radio

- Large on/off/volume button is very good.
- Preset buttons are large and clear; their positioning along the bottom of the unit is very good.
- Rocker seek button is satisfactory, good size and well located.
- Menu button a little small and awkward, also does not react enough when operated – could be more sensitive.
- News/TA buttons are well labelled and easy to operate.
- Pressing tape button for autoreverse function is a little unconventional, but a good way of saving buttons.
- Excellent idea to maintain FF/RWD buttons regardless of which side of the tape is playing.
- CD, AM/FM and Dolby buttons are well labelled and easy to use.
- Eject button is clear, easy to use and well positioned in relation to cassette door.

- Very good button consistency – all buttons have uniform size and labelling.
- PTY function is not very good; allocating generic titles to stations does not work very well.
- Display is well positioned and easy to read – informative and clear.
- RDS functions are a little obscure – required to read manual before initial operation.

Sharp radio

- On/off/volume control is a tad small and awkward, combined with difficult balance control.
- Push-button operation would be more satisfactory for on/off, as volume stays at preferred level.
- Fader control is particularly small and awkward.
- Both of the above points are related to the fact that a single button location has multiple functions – this is too complex.
- Treble and bass controls also difficult and stiff, although these functions are rarely adjusted once set.
- Station preset buttons are satisfactory – quite large and clear.
- Band selector button and FM mono/stereo button should not have two functions on each button – could cause confusion if wrong function occurs. These buttons are the only buttons on the radio which are not self-explanatory – the user must consult the manual to discover their function.
- Tuning seek and tuning scan buttons are easier to understand and use, although there are still two functions on the same button. These are probably used more than the aforementioned buttons.
- Cassette FF, RWD and eject buttons are self-explanatory; they have the accepted style for car radio designs. FF and RWD buttons could be a little larger.
- Autoreverse function is not so obvious, although it is an accepted standard (pressing FF and RWD buttons simultaneously).
- Illumination – is daytime/night-time illumination satisfactory? A dimmer control would probably aid matters.

Heuristics flowchart

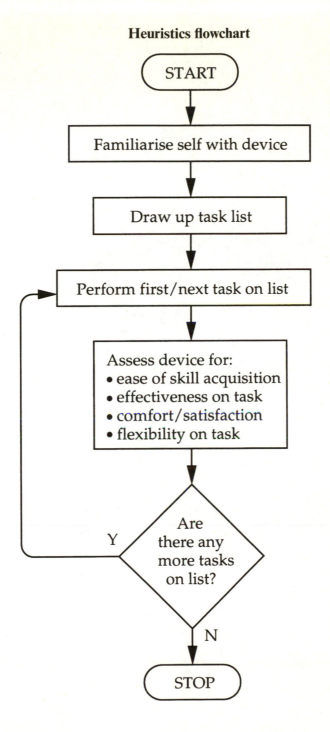

Section 3

UTILITY ANALYSIS

To determine the relative benefits of these methods when applied in the field, a utility analysis equation has been derived to assign an approximate financial value to each method. The equation is based on the accuracy of each method, the cost of retooling, i.e. 'things gone wrong' (TGW), and the cost in person hours of using the technique.

Accuracy

From a validation study, data have been acquired on inter-analyst reliability (how consistently different people use the method), intra-analyst reliability (how consistenly the same person uses the method on different occasions), and validity (how well the method predicts what it is supposed to). The coefficients of each of these variables lie between 0 and 1, so a simple multiplication will provide information on the overall accuracy of the method (which will also be between 0 and 1). That is, given a random analyst at any point in time, how well can they be expected to perform with a given method? Accuracy can therefore be summarised thus:

accuracy = inter-analyst reliability × intra-analyst reliability × validity

TGW Costs

The study was based on an in-car stereo system, so the retooling figures we refer to here are also based on a car radio. Of course, the analyst can substitute these figures for their own if they are interested in a different product. Retooling costs for a car radio can be between £3000 (for minor changes to the product line) and £150 000 (for complete retooling).

Assuming the accuracy of a method represents how much of these things gone wrong could be saved, multiplying the accuracy by the retooling costs will reveal how much money each method will yield:

savings = accuracy × retooling costs

Costs of using the method

But accuracy × retooling costs isn't the final figure, because there are costs involved in using the technique. If we assume an analyst is worth £50 per hour, each hour spent using the method will mean £50 less savings. Therefore our total utility for each method is

utility = (accuracy × retooling costs) − method costs

Substituting these equations into this one provides us with the final utility equation:

$$\text{utility} = (r_1 r_2 v C_t) - C_m$$

where r_1 = inter-analyst reliabilty
  r_2 = intra-analyst reliability
  v = validity

C_t = retooling costs
C_m = costs of using the method

Using the equation

There are four aspects of ergonomics which the methods attempt to predict: errors, performance times, usability and design. Most of the methods fit best into just one of these categories; the exception is observation, which can be used to predict both errors and performance times. The relationships between method and output may be summarised in a table.

Errors	Times	Usability	Design
PHEA	KLM	Checklists	Heuristics
Observation	Observation	Questionnaires	Link analysis
TAFEI		HTA	Layout analysis
		Repertory grids	
		Interviews	

Assume that the four areas account for an equal proportion of the retooling costs, and allow a similar-sized proportion for residual error; that makes five areas each accounting for 20% of the retooling costs. So, the first step in using the equation is to divide the retooling costs by 5.

Retooling costs will be specific to each situation, and this variable needs to be adjusted as appropriate. The rest of the variables for the equation may be summarised in a table.

Method	r_1	r_2	v	C_m (£)
KLM	0.754	0.916	0.769	112.5
Link analysis	0.286	0.830	0.764	104.2
Checklists	0.690	0.307	0.587	83.3
PHEA	0.551	0.392	0.614	241.7
Observation				
Errors	0.304	0.890	0.474	125
Times	0.209	0.623	0.729	125
Questionnaires	0.408	0.578	0.615	37.5
HTA	0.206	0.226	0.591	258.3
Repertory grids	0.157	0.562	0.533	112.5
Layout analysis	0.413	0.121	0.070	70.8
Interviews	0.334	0.449	0.466	283.4
Heuristics	0.064	0.471	0.476	62.5

The cost of the methods (C_m) is based on the time taken to analyse a car radio, and it includes training and practice time. And note that observation has two validity statistics associated with it, depending on whether errors or performance times are of primary concern. TAFEI is not included in the table, as it was not possible to collect accuracy data for this method. The remaining methods are presented in rank order of accuracy, starting with the most accurate.

Worked examples

Here are two examples of using the utility equation to demonstrate the payoff using a particular method on a car radio. The first demonstrates using the best method, KLM.

Step 1: Calculate retooling costs

A conservative estimate for retooling costs involved with a car radio could be set at £5000. KLM covers one area of ergonomics, performance times, so at best it can be expected to account for 20% of this, or £1000.

Step 2: Insert variables into the equation

$$\text{utility} = (r_1 r_2 v C_t) - C_m$$
$$U_{\text{KLM}} = (0.754 \times 0.916 \times 0.769 \times 1000) - 112.5$$
$$= 531.1 - 112.5$$
$$= £418.6$$

So, using KLM before commissioning this product could save about £420 on minor retooling costs.

Now let's try using heuristics and see how the figures compare.

$$\text{utility} = (r_1 r_2 v C_t) - C_m$$
$$U_{\text{Heuristics}} = (0.0644 \times 0.471 \times 0.476 \times 1000) - 62.5$$
$$= 14.4 - 62.5$$
$$= -£48.1$$

Here the costs of using the method outweigh the benefits. Of course, if the potential retooling costs were higher, the savings would be too, and this picture may well be different. In some cases, being restricted to one technique would be a disadvantage. How can utility be calculated for two or more techniques?

Using more than one method

The methods assess different aspects of ergonomics

If the chosen methods lie in separate categories then simply calculate the utility for each method separately and sum the amounts at the end. For a simple example, take the two methods already calculated. KLM assesses performance times, and heuristics is concerned with design. Of the £5000 total retooling costs, £1000 of this could be due to performance times, and a further £1000 due to design. So the total maximum potential saving is £2000.

$$\text{combined utility} = U_{\text{KLM}} + U_{\text{Heuristics}} = 418.6 + (-48.1) = £370.5$$

The methods assess the same aspect of ergonomics

This situation is slightly more complex. Because 20% of the retooling costs are allocated to each area, this proportion has to be shared somehow.

Assume the methods will be executed in order of accuracy, best first. Calculate the *savings* (not overall utility) for the first method. Then perform the utility analysis for the second method on the *remainder*. Sum the respective utilities at the end of this process and you have the overall utility for using the methods in combination.

Staying with the KLM example, let's say it is to be used with observation to predict performance times. The savings generated by KLM (before subtracting the costs of using the method) are £531.1, leaving £468.9 out of the original £1000. Now use the utility equation for observation on this £468.9 (beware to insert the correct validity statistic for observation predicting performance times):

FOR METHOD 1 (KLM):

$$\text{savings} = r_1 r_2 v C_t$$
$$S_{KLM} = 0.754 \times 0.916 \times 0.769 \times 1000$$
$$= 531.1$$
$$\text{remainder} = 1000 - 531.1 = 468.9$$

$$\text{utility} = (r_1 r_2 v C_t) - C_m$$
$$U_{KLM} = (0.754 \times 0.916 \times 0.769 \times 1000) - 112.5$$
$$= 531.1 - 112.5$$
$$= £418.6$$

FOR METHOD 2 (OBSERVATION):

$$\text{utility} = (r_1 r_2 v \times \text{remainder}) - C_m$$
$$U_{Obs} = (0.209 \times 0.623 \times 0.729 \times 468.9) - 125$$
$$= -£80.49$$

So the overall utility of using both KLM and observation is

$$418.6 + (-80.49) = £338.11$$

Summary

The utility equation described here is intended to provide an approximate insight into how much money each method is potentially worth to designers and engineers. It is purely a cost-benefit tool; it is not intended to be accurate enough for use in accounting. The reader should also be aware that the method costs (C_m) are based on analysing a car radio, so they may change with other devices. Retooling costs will also differ, so it is up to the analyst to substitute them accordingly. It is recommended that a conservative attitude is adopted for this.

However, these issues aside, the utility analysis can provide a tangible forecast about the usefulness of each method, albeit approximate. It may also aid in choosing between methods, for the relative advantages of one method over another may be more clearly defined in monetary terms.

Section 4

TRAINING, RELIABILITY, AND VALIDITY

This section reviews two studies we undertook for our EPSRC project on ergonomic methods. The first was a study of training people to use the methods and an assessment of their application to a device evaluation. The second was a study of the reliability and the validity of the methods. These data were derived from the comparison of the prediction from the first study to the performance of people using the device. Further details of the first study are reported by Stanton and Young (1998a).

Training people to use ergonomic methods

Very little has been written on training people to use ergonomic methods, as noted by Stanton and Stevenage (1998). In order to evaluate the ease with which people are able to acquire ergonomic methods, we conducted a study into the training and application of each method by novice analysts. In the first week, participants spent up to a maximum of 4 hours training per method, including time for practice. The training was based upon tutorial notes for training ergonomics methods developed by the authors. The training for each method consisted of an introduction to the main principles, an example of applying the method by case study, and the opportunity to practice applying the method on a simple device. In order to be consistent with other training regimes in ergonomic methods, the participants were split into small groups. In this way they were able to use each other for the interviews, observations, etc. At the end of the practice session, each group presented their results back to the whole group and experiences were shared. Timings were recorded for training and practice sessions. In the second and fourth weeks, participants applied each method in turn to the device under analysis. Timings were taken for each method, and subjective responses to the methods were recorded on a questionnaire on both occasions. A questionnaire was also used to gauge subjective reactions to the methods (Kirwan, 1992).

Although the data from the training and practice phase do not lend themselves to statistical analysis because they were taken for the group as a whole, they do present an interesting picture (Figure 4.1). These data seem to reinforce the reason for the popularity of questionnaires, interviews, observations, checklists and heuristics noted in the survey (Stanton and Young, 1998a) as they take relatively little time to learn when compared with HTA, PHEA and TAFEI. Perhaps it is surprising to see that link and layout analysis are not more popular, given that they are also relatively quick to train people in. Similarly, repertory grids and the KLM seem to be no more time-consuming to train people in than the focused interview. However, these techniques are rather more specialised in their output, like link and layout analysis.

The picture for application of the methods is rather similar, as Figure 4.2 shows. There is a statistically significant difference in the time taken to analyse a device using different approaches ($\chi^2 = 80.2094$, $p < 0.0001$ at trial 1; $\chi^2 = 72.3846$, $p < 0.0001$ at trial 2). We did not compute comparisons between individual methods because the non-parametric tests were not powerful enough to cope with the small sample size and the large number of ties in the data. Thus, as the overall ANOVA was statistically significant, it was deemed that a visual inspection of the results was sufficient. However, we did compare execution times for trial 1 and trial 2 within methods.

Statistical analysis shows a general reduction in execution times for seven of the methods:

- Heuristics ($Z = -2.6656$, $p < 0.01$)
- Checklists ($Z = -2.1974$, $p < 0.05$)
- HTA ($Z = -2.5205$, $p < 0.05$)
- PHEA ($Z = -2.3805$, $p < 0.05$)
- TAFEI ($Z = -2.3805$, $p < 0.05$)
- Repertory grids ($Z = -2.5205$, $p < 0.05$)
- KLM ($Z = -2.5205$, $p < 0.05$)

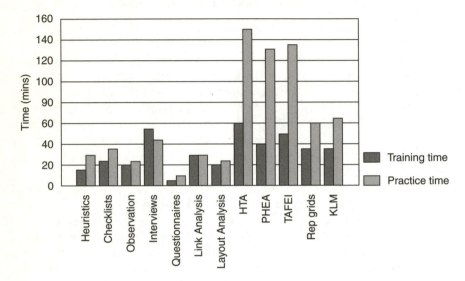

Figure 4.1 Training and practice times for ergonomic methods.

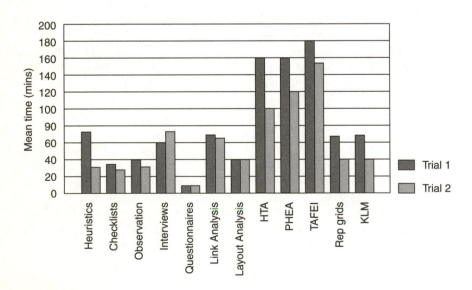

Figure 4.2 Average execution times for each method.

Figure 4.2 also shows that the popularity of questionnaires, observations and checklists is reinforced by their being relatively quick and flexible methods. The speed with which observations are conducted is perhaps counterintuitive, but the time taken in execution of the method is for a single participant only. And note that heuristics and interviews appear to take as long as link analysis, repertory grids and KLM, whereas layout analysis appears quicker. HTA and PHEA take approximately the same time as each other, but they are much more time-intensive than other methods. Besides that, PHEA requires the output of HTA, so it would require the time to conduct HTA plus the time to conduct PHEA if it were to be used in a situation where no HTA had been developed. Over half the participants failed to complete TAFEI within the time available, suggesting that it was the most time-consuming of the methods under test.

The subjective evaluation of the methods by the participants over the seven criteria (acceptability, auditability, comprehensiveness, consistency, resource usage, theoretical validity and usefulness) showed no effect of time. Only two statistically significant findings were found in the subjective evaluations; these findings were for the consistency of the methods (the degree to which the method is likely to produce the same result on successive occasions, analogous to test/retest reliability) and the resource usage (the amount of resources, usually time and effort, required to conduct the evaluation with a given method). Participants rate some methods as significantly less consistent than others ($\chi^2 = 39.6061$, $p < 0.0001$), as shown in Figure 4.3.

As Figure 4.3 shows, heuristics is rated as less consistent than any other method, whereas more structured techniques (e.g. checklists, HTA, PHEA and KLM) are rated as more consistent. It is ironic, but not surprising, that the highest rated method in terms of consistency was also rated as one of the least acceptable methods. Some methods were also rated as requiring significantly more resources than other methods ($\chi^2 = 37.6869$, $p < 0.0001$), as shown in Figure 4.4. This analysis seems to favour questionnaires, checklists, observation, repertory grids and KLM. HTA is obviously resource-intensive, as are PHEA, TAFEI, link analysis and interviews.

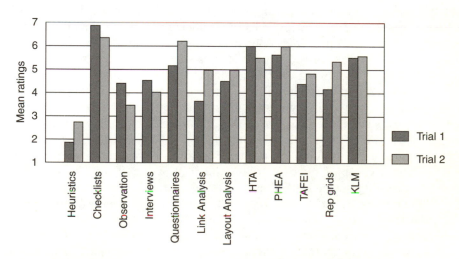

Figure 4.3 Consistency: average ratings on a seven-point scale (1 = poor, 7 = good).

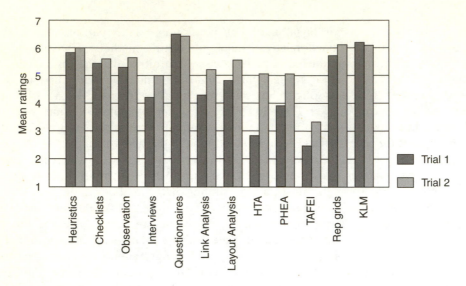

Figure 4.4 Resource usage: average ratings on a seven-point scale (1 = poor, 7 = good).

These analyses seem to suggest that some methods may be more acceptable than others because of the time required to learn to use them, the time they take to apply to an evaluation and the degree of consistency that they offer.

Reliability and validity

Not only do the methods have to be acceptable to the users, but they also have to work. The objective way to see whether the methods work is to assess their reliability and validity. If the methods can be demonstrated as reliable and valid, they may be used with confidence. The reliability of the methods was assessed in two ways. Intra-rater reliability was computed by comparing the output generated by each participant at trial 1 with the output at trial 2. Correlation coefficients, d', were computed to assess the stability of the measures. Inter-rater reliability was computed by looking at the homogeneity of the results of the analysts at trial 1 and at trial 2. In essence, the purpose of the validity study was to determine the extent to which the predictions were comparable to the actual behaviour of drivers when interacting with the radio-cassette. The data were analysed in different ways. First, intra-rater reliability was determined by using Pearson's correlation coefficient. This measures the degree of consistency of each rater at trial 1 compared with the same rater at trial 2. Second, inter-rater reliability was computed using the Kurtosis statistic. This looks at the degree of spread for the ratings with each group of raters at trial 1 and then at trial 2. Finally, validity was analysed by assessing the value of d' at trial 1 and at trial 2. This value is the combination of the *hit rate* and *false alarm rate*. The distinction is a significant one, because it is as important to predict true positives as it is to reject false positives.

The data analysis of reliability and validity are presented in Table 4.1. The reliability and validity data are presented together because the two concepts are interrelated. A method might be reliable (i.e. it might be stable across time and/or stable across

Table 4.1 Reliability and validity statistics for each method

Method	Inter-analyst reliability	Intra-analyst reliability	Validity
KLM	0.754	0.916	0.769
Link analysis	0.286	0.830	0.764
Checklists	0.690	0.307	0.587
PHEA	0.551	0.392	0.614
Observation			
Errors	0.304	0.890	0.474
Times	0.209	0.623	0.729
Questionnaires	0.408	0.578	0.615
HTA	0.206	0.226	0.591
Repertory grids	0.157	0.562	0.533
Layout analysis	0.413	0.121	0.070
Interviews	0.334	0.449	0.466
Heuristics	0.0644	0.471	0.476
TAFEI	–	–	0.506

Note: Kurtosis statistics have been transformed to give values between 0 and 1.

analysts) but it might not be valid (i.e. it might not predict behaviour). However, if a method is not reliable it cannot be valid. Therefore the relationship between reliability and validity can be described as unidirectional. The analysis for TAFEI is incomplete because not enough analysts completed the exercise within the four hours to enable the intra-rater and inter-rater reliability statistics to be computed.

In addressing intra-rater reliability, three of the methods achieved acceptable levels, denoted by the statistically significant correlations:

- Observation
- Link analysis
- KLM

This means that the analysts' predictions were stable across time.

Seven methods achieved acceptable levels of inter-rater reliability, as evidenced by the Kurtosis statistic (which is an indicator of how closely grouped the analysts' predictions were to each other), where a value greater than zero means that the data are steeper (therefore more tightly grouped) than the normal distribution curve, and a value less than zero means that the data are flatter (therefore more distributed) than the normal distribution curve. Ideally, values should be greater than zero to indicate greater agreement between raters. The more positive the value, the greater the degree of agreement. Generally speaking, the values improved between trial 1 and trial 2, suggesting the analysts were learning how to apply the techniques. Here are the methods that performed acceptably (at trial 2):

- Checklists
- Observation
- Questionnaires
- Layout analysis

- PHEA
- KLM

This means these methods showed an acceptable level of agreement between analysts. Finally, validity was computed from d' with the exception of observation, questionnaires and KLM (where Pearson's correlation coefficient was used). A value of $d' > 0.5$ was the acceptance criterion for the method (or a statistically significant correlation in the case of observation, questionnaires and KLM). Here are the methods that performed acceptably:

- Checklists
- Link analysis
- HTA
- PHEA
- TAFEI
- Repertory grids
- KLM

This means these methods seemed to capture some aspects of the performance of the participants engaged in the study of the radio-cassette. However, validation data cannot be interpreted independently of reliability data. Therefore only one of the methods performed at an acceptable level for all three criteria, KLM.

Relaxing these criteria a little would allow us to consider three more methods that performed at an acceptable level with respect to inter-rater reliability and predictive validity (with the proviso that the evidence suggests the methods are not stable within analysts for the first trial or over time for the second and third trials):

- Link analysis
- Checklists
- PHEA

Given that methods cannot be valid unless they are proven to be reliable, and there is little point in using methods that are reliable unless they are proven to be valid, we recommend that all the other methods are treated with caution until further studies have established their reliability and validity. But note that our data came from novice analysts who have only recently received training in the methods. Baber and Stanton (1996a) show much improved reliability and validity statistics for expert users of TAFEI and PHEA.

Summary of Findings

The general aim of this section was to assess the trainability, reliability and validity of the methods when presented to novices. The findings from these studies are summarised below. It is intended that these data can help guide the selection of methods, in addition to the information presented in Sections 1 and 2. The reliability and validity data are used as the basis for the utility analysis presented in Section 3.

Training people to use the method

Our studies have shown that initial training and practice time varies from method to method. Questionnaires are undoubtedly the quickest to train and practice whereas HTA, PHEA and TAFEI undoubtedly take the longest time of the methods we evaluated. Table 4.2 provides a rough guide to training times. This is the first study conducted, so exact time values must be treated with caution. But they do provide an approximation of the relative differences between the methods.

As this study trained participants in small groups, it was not possible to carry out individual analyses of performance to some predetermined criterion. This has obvious methodological limitations for the research, but we accept these limitations within the applied nature of the research project. These issues should perhaps be addressed in future research.

Time taken to apply the method to evaluate a device

Application times varied between methods. The questionnaire was undoubtedly the quickest to apply whereas HTA, PHEA and TAFEI took longer to apply in the device evaluation study. Comparing Tables 4.2 and 4.3, there are only two differences: heuristics and link analysis took longer to apply than to train and practice.

Relative benefits of methods

In assessing the relative benefits of these methods, we can consider the applicability of the approaches and the training and application times (which would favour the application of the questionnaire as a quick approach). In addition we assessed the subjective evaluation of the people who used the methods in our study. A survey undertaken by Stanton and Young (1998a) suggests that professional ergonomists prefer to use checklists and interviews. Checklists were rated as the most consistent method by the people in our training study, and questionnaires were rated as the least resource-intensive together with KLM.

Table 4.2 Combined training and practice times

Time	Methods
Over 2 hours	HTA, PHEA, TAFEI
1 to 2 hours	Interviews, repertory grids, KLM
30 to 60 minutes	Heuristics, checklists, observations, link analysis, layout analysis
Less than 30 minutes	Questionnaires

Table 4.3 Application times

Time	Methods
Over 2 hours	HTA, PHEA and TAFEI
1 to 2 hours	Heuristics, interviews, repertory grids, KLM
30 to 60 minutes	Checklists, observations, layout analysis
Less than 30 minutes	Questionnaires

115

Reliability and validity

In addressing the relibility and validity, we found that methods in our study did differ considerably. We suggest that newly trained novices perform better with some methods than others, as shown by the rank ordering in Figure 4.5; darker shading represents a better performance on that criterion. KLM performed best in our test. The next grouping was for link analysis, checklists and PHEA. The third grouping was the rest of the methods; we urge caution when they are used by novice analysts.

Conclusions

In conclusion, there is clearly little reported evidence in the literature on reliability or validity of ergonomic methods. This was confirmed by the survey undertaken by Stanton and Young (1998a). The detailed review of ergonomic methods reported in Section 2 leads to a greater insight into the demands and outputs of the methods under scrutiny. The training study indicated that link analysis, layout analysis, repertory grids

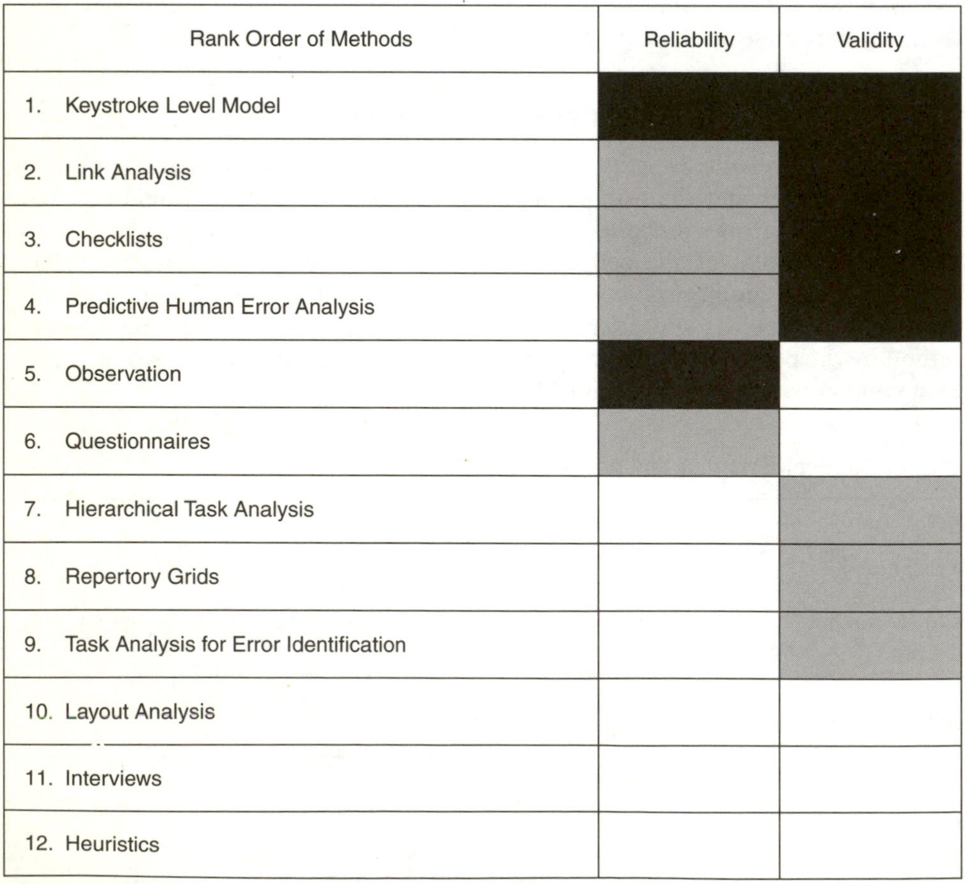

Rank Order of Methods	Reliability	Validity
1. Keystroke Level Model		
2. Link Analysis		
3. Checklists		
4. Predictive Human Error Analysis		
5. Observation		
6. Questionnaires		
7. Hierarchical Task Analysis		
8. Repertory Grids		
9. Task Analysis for Error Identification		
10. Layout Analysis		
11. Interviews		
12. Heuristics		

Figure 4.5 Rank order for reliability and validity (darker shading indicates better performance).

and KLM appear favourable when compared with more commonly used methods. The study of reliability and validity favours KLM, link analysis, checklists and PHEA. It is not by chance that the top-performing methods in terms of reliability and validity concentrate on very narrow aspects of performance (i.e. the execution of action and characteristics of the device; see the seven-stage model in Section 1).

Generally speaking, the broader the scope of the analysis, the more difficult it is to get favourable reliability and validity statistics. But this does not negate the analysis. We are arguing that analysts should be aware of the potential power of the method before they use it, rather than proposing that they should not use it. We also suggest that ergonomists and designers would be well served by exploring the benefits of other methods rather than always relying upon three or four favourite approaches. However, it is an important goal of future research to further establish the reliability and validity of ergonomic methods in different contexts.

A study of expert users would be useful as the data might be very different; a study of devices with greater and lesser complexity would be useful too. Analyst expertise and device complexity are likely to interact. Stanton and Stevenage (1998) found that novices performed significantly better than those in this study when analysing a much simpler device. And Baber and Stanton (1996a) showed that expert analysts performed significantly better than those in this study when analysing a much more complex device. The research debate is likely to continue for some time. It has taken researchers in the field of personnel selection some 40 years to reach a general consensus of opinion about the performance of their methods. We are likely to be well into the next millennium before a similar status is achieved for ergonomic methods.

Mini bibliography

Baber, C. and Stanton, N. A. (1996a) 'Human error identification techniques applied to public technology: predictions compared with observed use', *Applied Ergonomics*, **27**, 2, pp. 119–31.

Kirwan, B. (1992) 'Human error identification in human reliability assessment – part 2', *Applied Ergonomics*, **23**, pp. 371–81.

Stanton, N. A. and Stevenage, S. (1998) 'Learning to predict human error: issues of reliability, validity and acceptability', *Ergonomics*, **41**, 11, pp. 1737–56.

Stanton, N. A. and Young, M.S. (1998a) 'Is utility in the mind of the beholder? A study of ergonomics methods', *Applied Ergonomics*, **29**, 1, pp. 41–54.

Stanton, N. A. and Young, M.S. (1999b) 'What price erogonomics?', *Nature*, **399**, 197–8.

EPILOGUE

Utility Analysis

One of our critics pointed out that we do not need to factor reliability into the utility analysis equation as 'validity will already be attenuated by unreliability'. Our reason for including it was that we were trying to stay close to the style of utility analysis used for the evaluation of personnel selection methods. In this field the utility analysis formula considers the variation in performance of the potential personnel (called SDy) as well as the validity and cost of the methods in question. We have simply substituted reliability in place of SDy, as it represents the variability in the performance of people using the ergonomic methods. Although we accept they are not the same things, we thought we had captured the spirit of the analysis. We are not, however, going to object to people using the modified version of the formula with reliability removed:

$$(\text{validity} \times \text{retooling costs}) - \text{cost of using the method}$$

This has the following effect on the worked examples:

UTILITY OF KLM

$$U_{\text{KLM}} = (0.769 \times 1000) - 112.5$$
$$= £656.5$$

UTILITY OF HEURISTICS

$$U_{\text{Heuristics}} = (0.476 \times 1000) - 62.5$$
$$= £413.5$$

COMBINED UTILITY OF KLM AND HEURISTICS

$$U_{\text{KLM}} + U_{\text{Heuristics}} = 656.5 + 413.5 = £1070$$

COMBINED UTILITY OF KLM AND OBSERVATION

$$U_{KLM} + U_{Obs} = \{[(\text{retooling costs} - S_{KLM}) \times \text{validity}_{Obs}] - \text{cost}_{Obs}\} + U_{KLM}$$
$$= \{[(231) \times 0.729)] - 125\} + 656.5$$
$$= 43.4 + 656.5$$
$$= £699.9$$

These values represent a much more optimistic picture, but the relative differences between the methods is maintained (at least as far as the validity statistic is concerned). We accept this is not an exact science, but we are concerned that, when choosing ergonomic methods, the analyst has some degree of objectivity with which to support their judgement. Caution is recommended for the less reliable methods. The literature on utility has recently become concerned about the transparency of the analysis. We believe that our method is easy to use and it is very clear to see how the end result is derived.

Radio-cassettes

By now you are probably wondering which is the better radio-cassette, is it the Sharp or the Ford? Unfortunately, the results are not clear-cut. After comparing the outcome of the 12 methods, we have decided to call it a draw! This is not simply tactfulness or fencesitting on our part. Five methods favour the Sharp and five methods favour the Ford. We have left out HTA and link analysis data because they are inconclusive. HTA provides a purely descriptive account of task structure; it does not indicate which is a better task structure. The link analysis suggests better element groupings on the Ford, but there are fewer links on the Sharp. The success or failure of each design would depend upon the evaluation criteria developed a priori. Interestingly, it is all the quantitative methods (observation, questionnaire, PHEA, TAFEI and KLM) that suggest the Sharp is better and the qualitative methods (heuristics, checklists, interviews, layout analysis and repertory grids) that suggest the Ford is better. We think this provides an important lesson for designers: design evaluation should incorporate both quantitative and qualitative measures. With that final thought, we wish you well in designing the devices of the future.

GLOSSARY

Comprehensiveness	The broad scope or coverage of the method, e.g. does it make predictions about errors or performance times or both?
Consistency	How much of the method is logical and not contradictory.
Correlation coefficient	A measure of association between two variables. Ranges from -1 (perfect negative correlation) through 0 (no association) to $+1$ (perfect positive correlation).
Ecological validity	Real-world validity – validation through the presence of the study being based in an operational context rather than in an experimental laboratory.
Ergonomic methods	Methods and techniques which enable ergonomists to study, predict and understand human interaction with artifacts in the environment.
Ergonomics	The scientific study of human interaction with artifacts in the environment, e.g. the study of drivers using in-car devices while driving.
Hit rate	The proportion of hits to misses in a signal detection paradigm.
Inter-rater reliability	The consistency between many observers on a single measure.
Intra-rater reliability	The consistency of a single observer rating a measure over time.
Kurtosis	A measure of tightness or shape of distribution. Zero indicates a perfect normal distribution curve. Increasing negative values indicate a flattening distribution. Increasing positive values indicate a more peaked or closely grouped distribution. In this book it measures inter-rater reliability.
Methodology	A system of methods and principles to be used in ergonomics.
Participant pool	A list of people who have expressed a willingness to participate in psychological experiments and have registered a contact address.
Predictive validity	The accuracy of a method in predicting its associated variable, i.e. the degree to which the predictions are representative of observed behaviour.
Pro-forma	A set form or procedure.
Reliability	The consistency or stability of a measure, such as the degree to which a method will perform the same on separate occasions

	for the same person (intra-analyst reliability) and for different people (inter-analyst reliability).
Signal detection theory	A theoretical paradigm concerned with the factors that influence the ability to detect stimuli.
Theoretical validity	Validity based upon an underlying theory or model of human performance, e.g. a model of human cognition.
Usability	Ease of use or user friendliness, operationalised by Shackel into learnability, effectiveness, attitude and flexibility (LEAF).
User-centred design	Designing systems on the basis of a user requirements specification and an understanding of the capacities and limitations of the end-user population, sometimes involving user trials and user involvement in design stages.
Utility	Practical usefulness couched in financial terms.
Validation	Confirmation based upon evidence; *see* predictive validity and theoretical validity.
Validity	The accuracy of a measure to measure what it is supposed to measure.

BIBLIOGRAPHY

Annett, J., Duncan, K. D., Stammers, R. B. and Gray, M. J. (1971) *Task Analysis*, Department of Employment Training Information Paper 6.

Baber, C. (1996) Repertory grid theory and its application to product evaluation, in Jordan, P. W., Thomas, B., Weerdmeester, B. A. and McClelland, I. L. (eds) *Usability in Industry*, London: Taylor & Francis, pp. 157–65

Baber, C. and Stanton, N. A. (1994) 'Task analysis for error identification: a methodology for designing error-tolerant consumer products', *Ergonomics, 37*, pp. 1923–41.

Baber, C. and Stanton, N. A. (1996a) 'Human error identification techniques applied to public technology: predictions compared with observed use', *Applied Ergonomics, 27*,2, pp. 119–31.

Baber, C. and Stanton, N. A. (1996b) Observation as a technique for usability evaluations, in Jordan, P. W., Thomas, B., Weerdmeester, B. A. and McClelland, I. L. (eds) *Usability in Industry,* London: Taylor & Francis, pp. 85–94.

Baber, C., Hoyes, T. and Stanton, N. A. (1993) 'Comparison of GUIs and CUIs: appropriate ranges of actions and ease of use', *Displays, 14*, 4, pp. 207–15.

Brooke, J. (1996) SUS: a 'quick and dirty' usability scale, in Jordan, P. W., Thomas, B., Weerdmeester, B. A. and McClelland, I. L. (eds) *Usability Evaluation in Industry*, London: Taylor & Francis, pp. 189–94.

Butters, L. M. (1998) Consumer product evaluation: which method is best? A guide to human factors at Consumers' Association, in Stanton, N. A. (ed.) *Human Factors in Consumer Products,* London: Taylor & Francis, pp. 159–71.

Card, S. K., Moran, T. P. and Newell, A. (1983) *The Psychology of Human–Computer Interaction*, Hillsdale NJ: Lawrence Erlbaum.

Corlett, E. N., & Clarke, T. S. (1995). *The Ergonomics of Workspaces and Machines* (2nd edn), London: Taylor and Francis.

de Vries, G., Hartevelt, M. and Oosterholt, R. (1996) Private camera conversation: a new method for eliciting user responses, in Jordan, P. W., Thomas, B., Weerdmeester, B. A. and McClelland, I. L. (eds) *Usability Evaluation in industry,* London: Taylor & Francis, pp. 147–55.

Diaper, D. (1989) *Task Analysis in Human–Computer Interaction*, Chichester: Ellis Horwood.

Drury, C. G. (1990) Methods for direct observation of performance, in Wilson, J. and Corlett, E. N. (eds) *Evaluation of Human Work: A Practical Ergonomics Methodology,* 2nd edn, London: Taylor & Francis, pp. 45–68.

Dul, J. and Weerdmeester, B. (1993) *Ergonomics for Beginners: A Quick Reference Guide*, London: Taylor & Francis.

Easterby, R. (1984) Tasks, processes and display design, in Easterby, R. and Zwaga, H. (eds) *Information Design,* Chichester: John Wiley. pp. 19–36.

Eberts, R. (1997) Cognitive modeling, in Salvendy, G. (ed.) *Handbook of Human Factors and Ergonomics,* 2nd edn, New York: John Wiley, pp 1328–74.

Embrey, D. E. (1993) 'Quantitative and qualitative prediction of human error in safety assessments', *Institute of Chemical Engineers Symposium Series,* **130**, pp. 329–50.

Fransella, F. and Bannister, D. (1977) *A Manual for Repertory Grid Technique*, London: Academic Press.

Johnson, G. I. (1996) The usability checklist approach revisited, in Jordan, P. W., Thomas, B., Weerdmeester, B. A. and McClelland, I. L. (eds) *Usability Evaluation in Industry,* London: Taylor & Francis, pp. 179–88 .

Jordan, P. W. (1998) *An Introduction to Usability*, London: Taylor & Francis.

Jordan, P. W., Thomas, B., Weerdmeester, B. A. and McClelland, I. L. (1996) *Usability Evaluation in Industry*, London: Taylor & Francis.

Kelly, G. A. (1955), *The Psychology of Personal Constructs*, New York: Norton.

Kirakowski, J. (1996) The software usability measurement inventory: background and usage, in Jordan, P. W., Thomas, B., Weerdmeester, B. A. and McClelland, I. L. (eds) *Usability Evaluation in Industry*, London: Taylor & Francis, pp. 169–77.

Kirwan, B. (1990) Human reliability assessment, in Wilson, J. R. and Corlett, E. N. (eds) *Evaluation of Human Work: A Practical Ergonomics Methodology*, 2nd edn, London: Taylor & Francis, pp. 921–68.

Kirwan, B. (1992) 'Human error identification in human reliability assessment – part 2', *Applied Ergonomics,* **23**, pp. 371–81.

Kirwan, B. (1994) *A Guide to Practical Human Reliability Assessment*, London: Taylor & Francis.

Kirwan, B. & Ainsworth, L. K. (1992) *A Guide to Task Analysis*. London: Taylor & Francis.

Majoros, A. E. and Boyle, E. (1997) Maintainability, in Salvendy, G. (ed.) *Handbook of Human Factors and Ergonomics*, 2nd edn, New York: John Wiley, pp 1569–92.

Nielsen, J. and Molich, R. (1990) Heuristic evaluation of user interfaces, in Chew, J. C. and Whiteside, J. (eds) *Empowering People: CHI '90 Conference Proceedings*, Monterey CA: ACM Press, pp. 249–56.

Norman, D. A. (1988) *The Psychology of Everyday Things*, New York: Basic Books.

Oborne, D. J. (1982) *Ergonomics at Work*, Chichester: John Wiley.

Oosterwegel, A., and Wickland, R. A. (1995) *The Self in European and North American Culture: Development and Processes*, Dordrecht: Kluwer Academic.

Oppenheim, A. N. (1992) *Questionnaire Design, Interviewing and Attitude Measurement*, 2nd edn, London: Pinter.

Patrick, J., Spurgeon, P. and Shepherd, A. (1986) *A Guide to Task Analysis: Applications of Hierarchical Methods*, Occupational Services Publications.

Pejtersen, A. M., and Rasmussen, J. (1997) Effectiveness testing of complex systems, in Salvendy, G. (ed.) *Handbook of Human Factors and Ergonomics,* 2nd edn, New York: John Wiley, pp. 1514–42.

Ravden, S. J. and Johnson, G. I. (1989) *Evaluating Usability of Human–Computer Interfaces: A Practical Method*, Chichester: Ellis Horwood.

Salvendy, G. (1997) *Handbook of Human Factors and Ergonomics*, New York: John Wiley.

Salvendy, G. and Carayon, P. (1997) Data collection and evaluation of outcome measures, in Salvendy, G. (ed.) *Handbook of Human Factors and Ergonomics,* New York: John Wiley, pp. 1451–70.

Sanders, M. S. and McCormick, E. J. (1993) *Human Factors in Engineering and Design,* 7th edn, New York: McGraw-Hill.

Sanderson, P. M. and Fisher, C. (1997) Exploratory sequential data analysis: qualitative and quantitative handling of continuous observational data, in Salvendy, G. (ed.) *Handbook of Human Factors and Ergonomics,* New York: John Wiley, pp. 1471–1513.

Shepherd, A. (1989) Analysis and training in information technology tasks, in Diaper, D. (ed.) *Task Analysis for Human–Computer Interaction,* Chichester: Ellis Horwood, pp. 15–55.

Sinclair, M. A. (1990) Subjective assessment, in Wilson, J. R. and Corlett, E. N. (eds) *Evaluation of Human Work: A Practical Ergonomics Methodology*, 2nd edn, London: Taylor & Francis, pp. 69–100.

Stammers, R. B. (1996) Hierarchical task analysis: an overview, in Jordan, P. W., Thomas, B., Weerdmeester, B. A. and McClelland, I. L. (eds) *Usability Evaluation in Industry,* London: Taylor & Francis, pp. 207–13.

Stammers, R. B. and Shepherd, A. (1990) Task analysis, in Wilson, J. R. and Corlett, E. N. (eds) *Evaluation of Human Work: A Practical Ergonomics Methodology,* 2nd edn, London: Taylor & Francis, pp. 144–68.

Stanton, N. A. (1995) 'Analysing worker activity: a new approach to risk assessment?', *Health and Safety Bulletin,* **240**, Dec., pp. 9–11.

Stanton, N. A. (1998) *Human Factors in Consumer Products*, London: Taylor & Francis.

Stanton, N. A. and Baber, C. (1996a) 'A systems approach to human error identification', *Safety Science,* **22**, pp. 215–28.

Stanton, N. A. and Baber, C. (1996b) Task analysis for error identification: applying HEI to product design and evaluation, in Jordan, P. W., Thomas, B., Weerdmeester, B. A. and McClelland, I. L. (eds) *Usability evaluation in industry*, London: Taylor & Francis, pp. 215–24.

Stanton, N. A., and Baber, C. (1998) A systems analysis of consumer products, in Stanton, N. A. (ed.) *Human Factors in Consumer Products,* London: Taylor & Francis, pp. 75–90.

Stanton, N. A. and Stevenage, S. V. (1998) 'Learning to predict human error: issues of acceptability, reliability and validity', *Ergonomics,* **41**, 11, pp. 1737–56.

Stanton, N. A. and Young, M. S. (1997a) Validation: the best kept secret in ergonomics, in Harris, D. (ed.) *Engineering Psychology and Cognitive Ergonomics,* Vol.2, *Job Design and Product Design,* Aldershot: Ashgate, pp. 301–7.

Stanton, N. A. and Young, M. S. (1997b) Validation of ergonomics methods, in Seppala, P., Luopajavi, T., Nygard, C. and Mattila, M. (eds) *From Experience to Innovation*, London: Taylor & Francis.

Stanton, N. A. and Young, M. S. (1997c) Ergonomics methods in consumer product design and evaluation, in Stanton, N. A. (ed.) *Human Factors in Consumer Products,* London: Taylor & Francis, pp. 21–53.

Stanton, N. A. and Young, M. S. (1998a) 'Is utility in the mind of the beholder? A study of ergonomics methods', *Applied Ergonomics,* **29**, 1, pp. 41–54.

Stanton, N. A. and Young, M. S. (1998b) Ergonomics methods in consumer product design and evaluation, in Stanton, N. A. (ed.) *Human Factors in Consumer Products*, London: Taylor & Francis, pp. 21–53.

Stanton, N. A., and Young, M. S. (1999a) Utility analysis in cognitive ergonomics, in Harris, D. (ed.) *Engineering Psychology and Cognitive Ergonomics*, Vols 3/4, Aldershot: Ashgate.

Stanton, N. A. and Young, M. S. (1999b) 'What price ergonomics?', *Nature,* **399**, 197–8.

van Vianen, E., Thomas, B., and van Nieuwkasteele, M. (1996) A combined effort in the standardization of user interface testing, in Jordan, P. W., Thomas, B., Weerdmeester, B. A. and McClelland, I. L. (eds) *Usability Evaluation in Industry*, London: Taylor & Francis, pp. 7–17.

Wilson, J. R. and Corlett, E. N. (1995) *Evaluation of Human Work*, 2nd edn, London: Taylor & Francis.

Woodson, W. E., Tillman, B. and Tillman, P. (1992) *Human Factors Design Handbook,* 2nd edn, New York: McGraw-Hill.

Young, M. S. and Stanton, N. A. (1999) The interview as a usability tool, in Memon, A. and Bull, R. (eds) *Handbook on the Psychology of Interviewing,* Chichester: John Wiley.

INDEX